ENERGY

*Over all,
rocks, wood, and water,
brooded the spirit of repose,
and the silent energy of nature
stirred the soul of its inmost depths.*

—Thomas Cole, *Essay on American Scenery*, 1835

Written by Doug Wolven

Illustrated by Diane Valko

Edited by June Hetzel and Robert Windham

Carolea Williams,
Project Director

With special thanks to Dr. James Rynd, Biola University.

CTP ©1996, Creative Teaching Press, Inc., Cypress, CA 90630
Reproduction of activities in any manner for use in the classroom and not for commercial sale is permissible.
Reproduction of these materials for an entire school or for a school system is strictly prohibited.

Table of Contents

Introduction to Energy 3
   Setting the Stage 4
   The Scientific Method 6
   The Scientific Processes 7

Unit 1: Solar Energy 8
   Key Concepts *overhead reproducible* 9
   Solar Energy 10
   Solar Energy Storage 11
   Solar Energy Applications 13
   Extension Activities 17

Unit 2: Wind Energy 18
   Key Concepts *overhead reproducible* 19
   Wind Energy 20
   Wind Energy Storage 22
   Wind Energy Applications 26
   Extension Activities 27

Unit 3: Water Energy 28
   Key Concepts *overhead reproducible* 29
   Water Energy 30
   Water Energy Storage 31
   Water Energy Applications 36
   Extension Activities 37

Unit 4: Fossil Fuels 38
   Key Concepts *overhead reproducible* 39
   Fossil Fuels 40
   Fossil Fuel Storage 41
   Fossil Fuel Applications 45
   Extension Activities 46

Unit 5: Nuclear Energy 48
   Key Concepts *overhead reproducible* 49
   Nuclear Energy 50
   Nuclear Reactors 51
   Nuclear Energy Applications 55
   Extension Activities 57

Culminating Activities 58
Career Corner 59
Assessment 60
Performance Evaluation 61
Bibliography 62
Energy Resources 63
Glossary 64

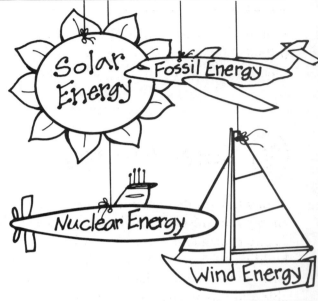

Introduction to Energy

Use this book as a guide to explore energy. Each unit opener includes illustrated key concepts followed by activities and experiments for meaningful, hands-on learning. Students practice the scientific method and processes as they work through the following sections:

Solar Energy

Students learn that solar energy is radiant energy that travels through space to the earth as electromagnetic waves. Activities include building solar cookers and fresh-water solar stills. Students learn that solar energy is renewable and explore ways for efficient energy harnessing through solar cells.

Wind Energy

Students learn that wind energy is a safe, renewable, clean source of energy derived from the natural movement of the air. They also learn that wind energy is an indirect source of solar energy. Students study the role of wind prospectors and use the Pythagorean theorem to determine wind velocity. They also build inclinometers and study the function of windmills in the conversion of wind energy to mechanical and electrical energy.

Water Energy

Students learn that water energy is another safe, clean source of energy and examine the conversion of energy from falling water to mechanical and electrical energy. Students study how dams and reservoirs work by building water pipes and experimenting with the effects of water pressure. They also examine two types of hydraulic turbines and build models of waterwheels, turbines, and generators.

Fossil Fuels

Students learn that fossil fuels are limited and are created from the remains of plants and animals exposed to extreme heat and pressure over an extended period of time. Students research fossil fuel consumption and predict future needs. Students make coal using the carbonization process and study fuel storage and application, including external and internal combustion engines.

Nuclear Energy

Students look at the atomic structure of matter and determine which elements have the potential to release nuclear energy through fission or fusion. Students study the storage and application of nuclear energy and debate the pros and cons of environmental impact.

Setting the Stage

Energy is in high demand in our world today. With fossil fuels nearing depletion, the world is searching for clean, efficient, renewable sources of energy. The following ideas set the stage for classroom investigations into the everyday concepts of energy. These activities provide a starting point from which students can research, record, organize, and evaluate their learning.

Science Journals

Staple lined paper into folded construction paper covers to create science journals. On the first page, have students create a knowledge chart by listing anything they know about energy. On the next page, have them record any thoughts, ideas, or questions they might have about energy. On subsequent pages, they can record the results of their experiments, ideas for further study, and responses to journal questions. Provide frequent opportunities for students to share journal entries.

Correspondence

Encourage students to write for information about energy to places such as:

Conservation and Renewable Energy Inquiry and Referral Service
P.O. Box 8900
Silver Springs, MD 20907

American Wind Energy Association
777 North Capitol Street NE
Washington, DC 20002-4226

Efficiency and Alternative Energy Technology Board Department of Energy, Mines, and Resources
580 Booth Street, 7th Floor
Ottawa, Ontario K1A OE4

Canadian Wind Energy Association
44A Clarey Avenue
Ottawa, Ontario K1S 2R7

For additional addresses, check your local yellow pages, the business index at your local library, or refer to the resource list on page 63.

Learning Center

Designate an area of the classroom as your Energy Research Center where students can work independently during their free time. Invite them to research various discoveries and inventions related to energy. Include reference books and posters of historic scientists for inspiration and a variety of materials such as paper, paper clips, straws, charcoal and colored pencils, paints, and markers. Provide shelf space for student inventions and models. Use bulletin boards to display student sketches and artwork related to energy.

The History of Energy

Provide a visual overview of energy history by posting a time line and a world map where students can chart locations of important discoveries. Small flags on pins can label the location, name, date, and event. Below the map, provide space for students to post index cards providing more detail.

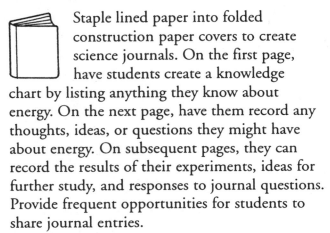

"FIRSTS" in Energy

1556 — Germany
Georgius Agricola wrote about water-wheel-driven mining pumps.

1764 — England
James Watt invented the separate condenser steam engine.

1840 — England
James Joule proved that heat, a form of energy, is created in an electrical circuit.

1873 — Scotland
James C. Maxwell formulated the theory of electromagnetic radiation.

Setting the Stage

Famous Faces in Energy

Invite students to research a famous figure in the field of energy, directing them to use a class chart to record this individual's role or contributions. Have students choose from scientists such as Faraday, Volta, and Einstein. Invite students to write and illustrate a biography based on the life of one of these famous scientists. Bind student books and include them in your classroom library.

Famous Faces in Energy

Person	Heritage	Dates	Role
Alessandro Volta	Italian	1745–1827	physicist
Michael Faraday	English	1791–1867	chemist and physicist
Thomas Edison	American (USA)	1847–1931	inventor
Ernest Rutherford	British (New Zealand born)	1871–1937	physicist
Albert Einstein	German-Swiss-American (USA)	1879–1955	mathematical physicist
Niels Bohr	Danish	1885–1962	physicist

The Scientific Method

Throughout each unit, hands-on experiments involve students in the steps of the scientific method. *Science Challenge* questions give students a chance to plan additional experiments. As students raise questions, they can use the Scientific Method reproducible on page 6 to help them design their own experiments.

The Scientific Processes

The eight scientific processes are integrated throughout the activities and experiments in each unit. Students can use the reproducible on page 7 to record and reflect on how they used the processes to reach their scientific conclusions.

Energy Library

Use the bibliography of nonfiction and fiction resources on pages 62–63 to begin your energy library. Add student-authored books to the library as you progress through each unit.

Name:

The Scientific Method

Question: Based on your observations, identify the problem you wish to explore. Then ask a clear, specific, testable question.
_____
_____

Hypothesis: Make a prediction or your best guess as to the solution or answer to your question. _____
_____
_____

Procedure: Plan the materials you will need and the steps you will take to test your hypothesis. _____
_____
_____
_____
_____
_____
_____
_____

Results and Conclusions: Record the results of your experiment.
_____
_____
_____
_____
_____
_____
_____
_____
_____
_____
_____
_____

Name: _____

The Scientific Processes

Observing: What was the most important thing you observed during the science experiment? _____
_____

Communicating: How did you communicate what you learned?
_____
_____

Comparing: How were the results of your experiment the same or different from the results of your classmates? _____
_____

Ordering: Did you notice any patterns in your data? What in your experiment reflected order in nature? _____
_____
_____

Categorizing: What was the most important thing you observed during the science experiment? _____
_____

Relating: What did you learn from this experiment that was related to a fact you previously knew? _____
_____

Inferring: Based upon what you learned in your experiment, are there other conclusions you might infer that are not direct observations?
_____
_____

Applying: How might you use what you learned to invent something practical to help our world? to explain how something else works?
_____
_____

Creative Teaching Press/*Energy*

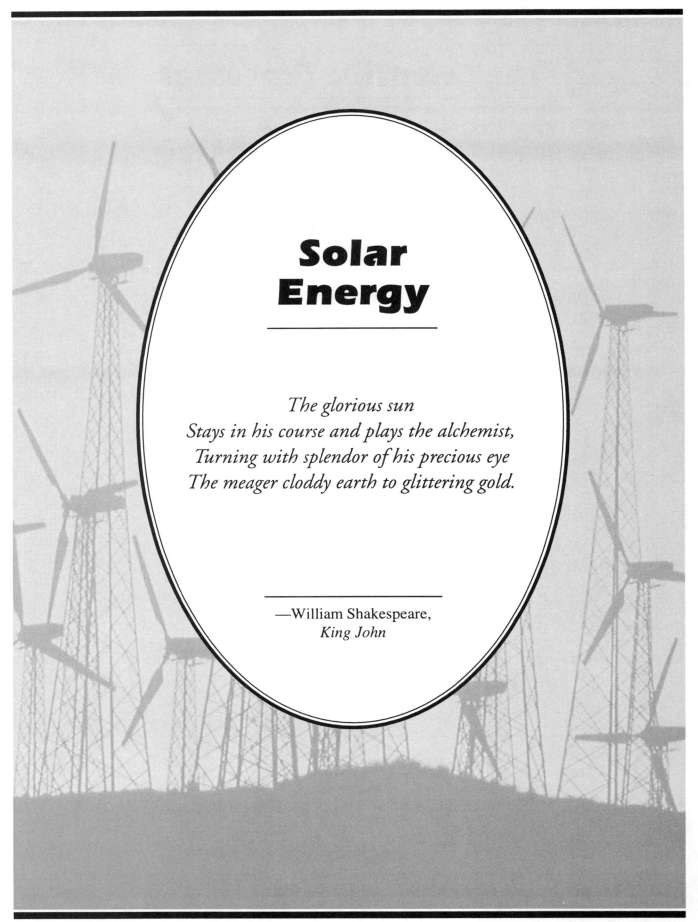

Solar Energy

*The glorious sun
Stays in his course and plays the alchemist,
Turning with splendor of his precious eye
The meager cloddy earth to glittering gold.*

—William Shakespeare,
King John

Key Concepts of Solar Energy

Solar Energy—
radiant energy from the sun

Solar Energy Storage—
energy stored in fossil fuels, plants, and solar cells

Solar Energy Applications—
energy for tools and machines such as electrical appliances, calculators, and water heaters

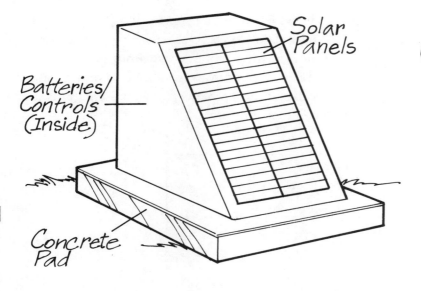

Solar Energy

Solar energy is **radiant energy** or **radiation.** Radiant energy is **electromagnetic energy** that travels through space as waves. Every day, the sun sends to Earth 400 million times more energy than we use. Energy from the sun strikes the earth continually. Without the constant flow of energy from the sun, the earth's temperature would be 200 degrees centigrade below zero and no life would exist. The sun is a free, renewable source of energy, but the equipment needed to convert solar energy to electricity can be very expensive.

Hot Surfaces
(Science Experiment)

Have you ever noticed how hot asphalt surfaces become on a summer day? Have students label identically-sized cups 1–5. Fill Cup 1 with water, Cup 2 with dry sand, Cup 3 with wet sand, Cup 4 with dry dirt, and Cup 5 with wet dirt. Insert a thermometer into each cup and place the cups in the sun for several hours. Have students record the temperatures of the cups every 15 minutes. (Note: Dry, dark surfaces absorb more heat than light, wet surfaces. The wet sand should remain coolest while the dry dirt gains the most heat.)

Electromagnetic Spectrum
(Science Demonstration/Research)

The electromagnetic spectrum includes radio waves, infrared waves, visible light, ultraviolet light, microwaves, X rays, and gamma waves. Humans can see wavelengths that range from approximately 400 to 700 nanometers. Invite students to use prisms to observe the visible spectrum (rainbow). Infrared light is radiation with wavelengths longer than visible light, but shorter than radio waves. Infrared rays are commonly called heat waves. Ultraviolet rays have wavelengths shorter than visible light, but longer than X rays. Ultraviolet rays can cause sunburn. Invite cooperative groups to study different types of waves in the electromagnetic spectrum and prepare creative presentations to share their findings.

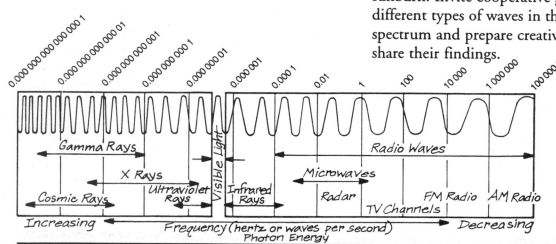

Solar Energy Storage

Solar energy is stored in plants, fossil fuels, and solar batteries. Green plants, natural storehouses of solar energy, use solar energy to make food through a process called **photosynthesis.** The food energy is stored in the plant as sugars and starch. These sugars contain carbon which can be converted to energy when people and animals digest them. **Fossil fuels** are another example of natural solar energy storage. The decayed plant and animal matter originally received their energy from the sun. Fossil fuels (oil, coal, and natural gas) provide 95% of the world's energy. **Power plants** can also store electricity in batteries similar to the type found in most cars.

Natural Storage Units
(Writing/Art)

Trees store energy in the cambium layer beneath their bark as well as in their fruit. Plants store energy in leaves and roots. This energy is passed on to animals through consumption. Ask students to suggest types of fruits, leaves, and roots they eat. Have students make a chart to record and illustrate the more popular sources of "stored energy."

Sources of Stored Energy

Roots	Leaves	Fruit
carrots	lettuce	apples
radishes	spinach	bananas
turnips		oranges

Power Plants
(Problem Solving/Creative Writing)

Power plants are hydroelectric-, thermal/fossil fuel-, or nuclear-powered facilities that generate electricity. Power plants consist of fuel-storage units and energy-release mechanisms such as dams, boilers, and reactors. Power plants also use turbines, generators, transformers, and transmission or electrical distribution systems. Challenge students to design a power plant, write an adventure that takes place in a power plant, or write a story about what their lives would be like without electricity generated by power plants. In each case, encourage students to use related terms such as *dam, boiler,* and *reactor.*

Solar Energy Storage

Solar Collector
(Science Experiment)

Solar collectors convert solar energy into heat energy. This energy can be harnessed to heat our homes. Invite students to make their own solar collectors by following these steps:

1. Fill a shoe box half full with crumpled newspaper.
2. Spray black paint on the newspaper and the inside of the box.
3. Punch a hole wide enough to fit plastic tubing through each end of the box.
4. Arrange tubing as shown.
5. Tape tubing in place.
6. Place a funnel at one end and a glass beaker at the opposite end.
7. Cover the box tightly with plastic wrap.
8. Place the box in direct sunlight, resting one end on a thin book to slightly tilt the box. The funnel should be at the high end.
9. Pour water into the funnel. Allow it to flow through the tube and collect in the beaker.
10. Repeat step 9 ten to twenty times, recording the water temperature each time.

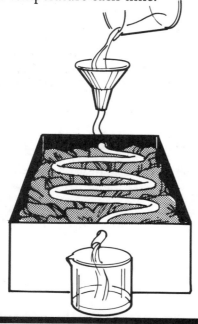

Potential Energy
(Science Experiment/Graphing)

Potential energy is energy waiting to be used for work or motion. It is the capacity an object has for doing work. An object possesses potential energy because of its position or condition. A pencil resting on the edge of a desk, about to fall to the floor, has potential energy because of its position. A stretched rubber band, ready to snap back, has potential energy because of its condition. Potential energy is stored in a bow when the arrow is set and drawn back on the bow string. When the arrow is released, the potential energy becomes kinetic (moving) energy. Kinetic energy can drive the released arrow to over 100 miles per hour.

Hold a new tennis ball above the ground. Explain to students that in this position, the ball has potential energy. Release the tennis ball and watch it bounce. The bouncing ball has kinetic (moving) energy. Have students predict the relationship between the height of the ball before it is dropped and the height of the first bounce. Then have students create a chart that compares the height of release with the height of the resulting bounce.

Height of first bounce in inches

	0	6	12	18	24	30	36	42
6								
5								
4								
3								
2								
1								

Height of drop in feet

Have students respond to these questions in their journals: *What is the relationship between the height of the tennis ball and the amount of stored energy? What is the difference between potential and kinetic energy? How do plants, fossil fuels, and solar cells store potential energy?*

Solar Energy Applications

Earth absorbs approximately 35% of the energy that reaches it from the sun. Most of this energy is spent evaporating moisture into clouds. Photosynthesis converts some solar energy into organic chemical energy. Scientists are looking for ways to use solar energy in the form of liquid heat storage and more efficient generation of electricity through **solar cells.** The light and night (diurnal) cycle, seasonal and climatic variations, and the currently cheaper energy from fossil fuels hinder the effective use of solar energy. Still expensive to manufacture, many solar cells are required to collect usable amounts of solar energy.

Satellites
(Model Building/Math)

Man-made satellites orbit the earth, continually bombarded by solar rays. Satellites use the sun's energy to run electric circuits. Have students use CD-ROMs, encyclopedias, and library books to find pictures of satellites with solar panels. Have students examine these photos and describe what different types of solar panels have in common. (They are very large, usually flat, exposed to the sun, and black.)
Have students build models of solar panels using toothpicks, glue, and black construction paper. Challenge students to build models to scale and calculate the surface area of their solar panels.

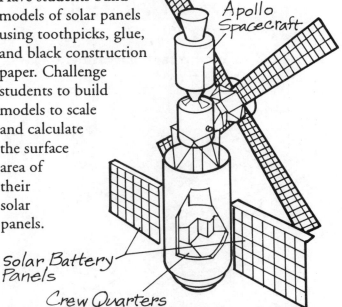

Solar-Powered Calculators
(Science Demonstration)

Ask students to enter several digits on a solar calculator. Have students cover several of the solar cells at the top of the calculator and observe the effect on the numbers. Ask students to write a short paragraph describing their observations, explaining what happened.

Solar Cooker
(Science Demonstration)

Invite students to build the solar cooker described on page 14. Then ask them to complete the experiment on pages 15 and 16 to harness the sun's rays to cook hot dogs.

Creative Teaching Press/*Energy*

Name:

Solar Cooker

Caution: Use saw only with adult supervision.

Materials
- 2 sealed sandwich bags (3/4 full of rice, beans, or sand)
- large, round oatmeal box with lid
- white glue
- cereal box top
- 2 paint stirrers
- small saw
- aluminum foil
- scissors
- stapler
- hole punch

Procedure:

1. Make two cuts, just over halfway around the oatmeal box, about two centimeters from the top and two centimeters from the bottom. Make sure the top and bottom cuts are parallel.

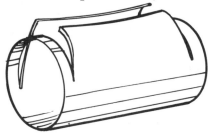

2. Slit the box from the middle of the top cut to the middle of the bottom cut. The sides of the box should open like the bay doors on the space shuttle.

3. Using the saw, cut the paint stirrers to 14 cm and staple them outside the "doors" to flatten and expose the doors to the sun. Start the staples from the inside.

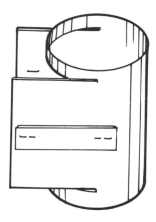

4. Cover the inside of the box with aluminum foil, shiny side up. Avoid crinkling the foil as much as possible.

5. Cut two cardboard strips (3 cm x 10 cm) from the cereal box top. Punch three holes in the strips about 1.5 cm apart.

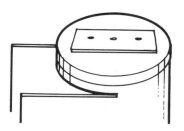

6. Glue the cardboard strips to the outside of the lid and bottom of the cooker. (See illustration for proper alignment.) Punch additional holes in both ends of the cooker that line up with the cardboard strips.

7. Rest the cooker on two bags of rice, beans, or sand so it faces the sun.

8. Use the solar cooker in Experiment #1 on pages 15 and 16.

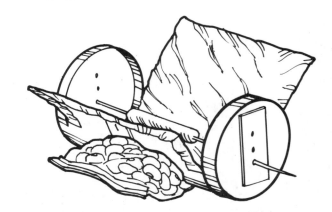

Name: _____

Solar Cooker
Experiment 1

EXPERIMENT 1

Question: How does a solar cooker maximize sun rays to cook food?

Hypothesis: _____
_____
_____

Materials
- solar cooker (from page 14)
- skewer (wood or metal)
- 4 hot dogs
- plate
- clock or timer

Procedure:

Caution: Use care when inserting skewer into hot dog.

Step 1

Run the skewer through a hot dog. Insert the skewer into the lowest set of holes. Keep the cooker pointed toward the sun to efficiently use the solar rays.

Step 2

Place another hot dog on a plate. Set the plate in the sun next to the cooker. This hot dog serves as the control (source of comparison), and helps determine if a hot dog left in the sun will cook without the aid of a cooker.

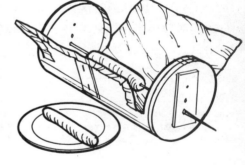

Step 3

Time the experiment. When the experimental hot dog in the cooker is done, it will have beads of moisture on its surface.

Step 4

Place another hot dog in the cooker using the middle holes. Time the experiment to see if the second hot dog cooks slower or faster than the first experimental hot dog.

Step 5

Try the experiment a third time using the top holes.

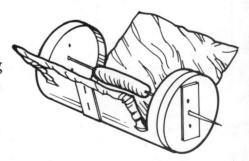

Creative Teaching Press/*Energy*   15

Name: _____

Solar Cooker
Experiment 1

EXPERIMENT 1

Results and Conclusions:

1. How long did it take the first hot dog to cook? _____ the second hot dog? _____ the third? _____ Which position was most effective? Why? _____
_____

2. Describe the condition of the control hot dog at the end of the same amount of time it took the experimental hot dogs to cook. _____
_____
_____

3. Describe the role of the foil in harnessing the sun's solar energy.
_____
_____

4. What part do you think air currents may have played in cooling the second hot dog? Why do you think commercially-sold solar ovens have glass covers? _____
_____
_____

5. How did the results of the experiment compare with your hypothesis?
_____
_____

6. Explain in some detail how you would improve/change the solar cooker if you were to rebuild it. Be sure to include a description of the materials and their effectiveness in harnessing solar rays. _____
_____
_____
_____

Science Challenge: Set up an experiment to test this question:
How can mirrors be used to cook food?
Write your question, hypothesis, procedure, and materials list on another sheet of paper. Then test the hypothesis and record your conclusions.

Creative Teaching Press/*Energy*

Extension Activities

Freshwater Solar Still
(Model Building/Writing)

Have students make freshwater solar stills to show how the sun's energy can be used to extract fresh water from air and soil. Each group of students needs 20" of plastic wrap, a shovel, an area of sand or soil in which they can dig a one-foot-deep hole, a cup to catch moisture, and rocks or bricks to hold the plastic wrap in place. Have students follow these steps:

1. Dig a hole one foot wide and one foot deep.
2. Place the cup in the bottom of the hole.
3. Place the plastic wrap over the hole so that it sags in the middle. The sag should be over the cup.
4. Secure the plastic wrap with rocks or bricks so it will not fall into the hole.

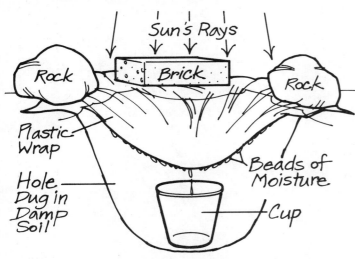

Sunlight penetrates the plastic and heats the soil. The moisture in the heated soil evaporates and rises up to the plastic. The air outside cools the surface of the plastic, causing the evaporated moisture to condense. As the water condenses, it runs down the inside surface of the plastic and drips into the cup. Upon completing this activity, have students illustrate and describe in their science journals how the freshwater solar stills work.

Greek Myths
(Literature/Writing)

The ancient Greeks invented stories called myths to help explain nature and the unknown. Just as energy can change form, so can gods, goddesses, and mortals as they transform themselves in these wonderful tales. Read a selection of Greek myths to the class and point out the sun and wind gods. Have students write their own myths, incorporating the power of the sun, wind, and water. Students can share their myths orally, just as the Greeks did.

Calories
(Science Demonstration)

The sun provides energy for plants to grow. Our bodies use plants for fuel. This fuel creates heat. The energy in food is measured in calories. A small calorie is the amount of energy necessary to raise the temperature of one gram of water one degree Celsius. Kilocalories, or large calories, are a measure of potential energy stored in food. Kilocalories store 1,000 times the energy of small calories.

Place 50 mL of water in a large test tube. Take the temperature of the water. Insert a marshmallow, sugar cube, or nut onto the blunt end of a needle. Stick the pointed end of the needle into a cork. Place the cork on a tabletop. Using a match or lighter, carefully light the food item on fire. Using a test tube holder, heat the water by placing the test tube over the flame. When the flame goes out, measure the temperature of the water. Subtract the first temperature reading from the second. This is the amount of heat energy your food substance emits. Repeat this procedure with different foods.

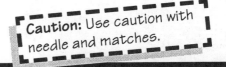

Wind Energy

*Loud wind, strong wind,
sweeping o'er the mountains,
Fresh wind, free wind,
blowing from the sea,
Pour forth thy vials like streams
from airy fountains,
Draughts of life to me.*

—Dinah Maria Mulock Craik,
North Wind

Key Concepts of Wind Energy

Wind Energy—
safe, renewable, non-polluting energy derived from natural air movement

Wind Energy Storage—
converted wind energy stored in power facility batteries for later use

Wind Energy Applications—
electricity for homes, mechanical power for pumping water and grinding grain, and power for sailing boats

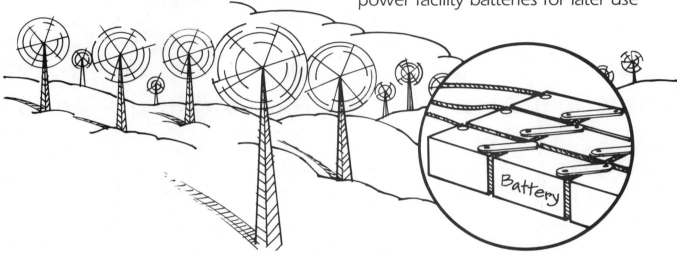

Wind Energy

Wind is an effect of solar energy. The sun's energy causes air to move, producing wind. Other factors affecting wind include atmospheric conditions, surface features of the earth, and the earth's rotation. A wind is named according to the compass direction from which it blows. For example, a north wind blows from the north. The wind's energy can be harnessed through the use of windmills. Although wind is free, it is not constant. Efficient use of wind energy involves placing **windmills** in windy locations. **Wind prospectors** are people who search for locations that experience consistent and strong wind flow.

Pinwheel Turbine
(Science Experiment)

Electricity can be generated by spinning large turbines. Have students use pinwheels as models of turbines to see how many ways they can make pinwheel turbines spin. Student activities can include blowing air over the pinwheels, holding them out as they ride bicycles, and running with the pinwheels.

Wind Tunnel
(Science Experiment)

To demonstrate the power of moving air, have students place two books of the same thickness about five inches apart, and lay a sheet of paper lengthwise across the books. Have students blow straight across under the paper and parallel to the table. The moving air creates lower pressure below the paper relative to the pressure above it, pulling the paper into the crevice. This demonstrates Bernoulli's Principle—moving air creates a low-pressure area or partial vacuum.

Blowing in the Wind
(Science Experiment)

Students can predict and measure wind speed by building their own anemometers, tools for measuring wind velocity.

1. Use a sewing needle to thread about 12 inches of heavy-duty thread through a Ping-Pong™ ball. Tie a knot at the end of the thread to secure the ball.
2. Glue the free end of the thread to the center guide (hole) of a protractor.
3. Hold the protractor perpendicular to the ground so the thread runs along the 90-degree mark.
4. Take the anemometer outside.
5. Record the degree line to which the string moves as the wind blows the ball. Keep a daily record. Use the conversion table to convert degrees to approximate kilometers per hour.

Angle	Wind Speed (km/h)	Angle	Wind Speed (km/h)
90	0	50	18
85	6	45	20
80	8	40	21
75	10	35	23
70	12	30	26
65	13	25	29
60	15	20	33
55	16		

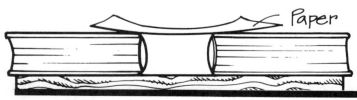

Wind Energy

Rising Air
(Science Demonstration/Writing)

Warm air rises. Demonstrate rising warm air using a small, empty glass soda bottle; a small, deflated balloon; a bucket of hot water; and a bucket of ice water. Follow these steps:

1. Place the empty soda bottle in the bucket of ice water for a few minutes.
2. Remove the bottle from the ice water and stretch the neck of the deflated balloon around the mouth of the cooled bottle. The balloon stays deflated.
3. Pass the bottle around to several students. As they hold the bottle, the cool air inside begins to warm, and the balloon starts to inflate as warm air molecules move farther apart.
4. Place the bottle back into the bucket of ice water. Observe the balloon deflate.
5. Ask students to predict what will happen. After the balloon deflates, place the bottle in the bucket of hot water. (The balloon expands.)
6. Explain to students that warm air molecules expanded to increase air pressure.

Have students respond to the following questions in their science journals: *Why does the balloon inflate? Why does the balloon deflate? How did the temperature of the water affect the air? Wind is moving air. Was wind created in the bottle? Which direction did the air current flow?*

Scale	Description	Wind mph
0	calm— smoke drifts up	less than 1
1	light air— smoke drifts with wind	1-3
2	light breeze— leaves rustle	4-7
3	gentle breeze— leaves move constantly	8-12
4	moderate breeze— branches move	13-18
5	fresh breeze— small trees sway	19-24
6	strong breeze— large branches move	25-31
7	moderate gale— whole trees move	32-38
8	fresh gale— twigs break	39-46
9	strong gale— slight damage to houses	47-54
10	whole gale— much damage to houses	55-63
11	storm— extensive damage	64-75
12	hurricane— extreme damage	more than 75

Beaufort's Wind Scale
(Meteorology/Math)

Windmills can face breezes too light to turn their blades or gales so powerful that blades can be damaged. Scientists must be careful where they place windmills so that neither of these situations occur. Sir Francis Beaufort created a scale which classifies wind velocity and its effects. This scale is accepted internationally and referred to on weather forecasts.

Have students monitor weather reports, noting wind speeds for seven consecutive days. Have students list the scale number and name of wind for each day's report. Ask students to note changes in speed or direction. Students can then determine the average daily wind velocity for the week.

Creative Teaching Press/*Energy*

Wind Energy Storage

Wind is safe, **renewable,** and **non-polluting.** Wind energy is free once the equipment for energy conversion is built and installed. Modern windmills turn generators to make electricity. Most wind power is used immediately, but for those times when the wind does not blow, wind energy (electricity) must be stored in lead-acid utility scale batteries. Usually the best places for wind machines are on mountains and near the seashore.

Batteries
(Technical Drawing/Writing)

Unused wind energy must be stored in batteries. Familiarize students with batteries by bringing in several samples. Have them find an illustration of a battery in an encyclopedia or book. Have students draw and label a battery diagram. Ask students to write an explanation of how they think a battery works.

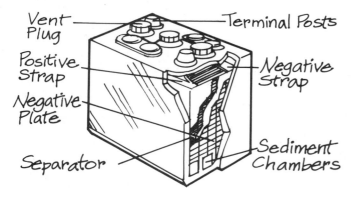

Wind Velocity
(Science Experiment)

In Experiment #2 on pages 24 and 25, students use an inclinometer (from page 23), kites, and the Pythagorean theorem to determine wind velocity.

Zinc and Penny Battery
(Science Experiment)

Help students understand how stored energy can be accessed in batteries. Have them make a simple battery from a scrap of zinc or tin, two 20-cm bell wire leads, a polished penny, and a piece of construction paper or cloth soaked in salt water. Use the galvanometer from page 27 to test this do-it-yourself battery. Have students follow these steps:

1. Cut a 35-mm square of tin or zinc.

Caution: Cut metal under adult supervision.

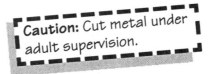

2. Sandwich the saltwater-soaked paper or cloth between the metal and the penny.
3. Touch one wire to the top and one wire to the bottom of the battery "sandwich."
4. Connect the other ends of the two wires to the galvanometer.
5. Observe the compass activity.

Name:

Inclinometer

Materials
- protractor pattern
- 1" x 4" x 24" pine board
- 1/4" x 24" dowel
- three 3/4" eye screws
- poster board
- glue

The height of objects can be measured using inclinometers. Inclinometers can also be used to determine wind altitude.

Procedure:

1. Insert an eye screw into each end of the dowel.

2. Place the board flat on a table. Center and insert an eye screw into the top of the board about one inch from the top edge.

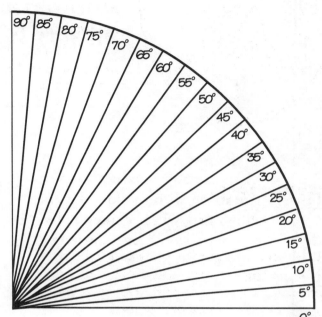

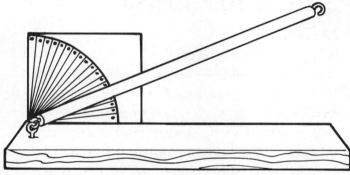

3. Join one end of the dowel to the board by interlocking the eye screws.

4. Glue the protractor pattern onto the poster board. Glue the poster board to the side of the pine board with the protractor pattern facing the dowel. The junction of the two interlocking eye screws forms an angle. The kite string from Experiment #2 on pages 24 and 25 will be held at this junction when the kite is aloft.

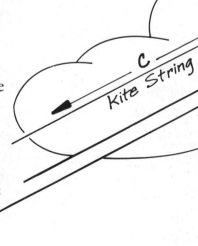

(Inclinometer not to scale.)

Creative Teaching Press/*Energy* 23

Name: _____

**Wind Velocity
Experiment 2**

EXPERIMENT 2

Question: How can kites help us understand how altitude affects wind velocity?

Hypothesis: _____

Materials
- ☐ paper kite
- ☐ 60 plus meters of string
- ☐ masking tape
- ☐ inclinometer (from page 23)
- ☐ measuring tape

Procedure:

Step 1

Mark the 60-meter point on the kite string with masking tape. This string will become the hypotenuse, side c, of the triangle. Multiply 60 by .95 to correct for the sag of the string (60 x .95 = 57).

Step 2

Fly your kite, releasing the string to the 60-meter tape mark.

Step 3

Place the inclinometer at the 60-meter mark on the string. Insert the kite string into the eye hook at the top of the inclinometer. This will lift the dowel, indicating the angle of the string. Be sure to hold the base of the inclinometer parallel to the ground.

Step 4

Measure the distance from the eye hook to the point directly below the kite. This distance is side a of the triangle. The point under the kite provides a 90° angle.

Step 5

Since you know sides c and a, you can now calculate side b of the triangle to give you the height of the kite, using the Pythagorean theorem: $a^2 + b^2 = c^2$ or $b^2 = c^2 - a^2$.

Step 6

Fly the kite at several different heights to compare wind strengths at various levels. Remember that your kite is pushed higher by stronger winds.

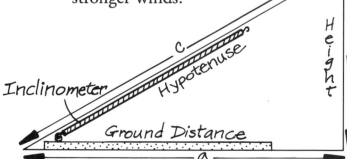

24

Creative Teaching Press/*Energy*

Name:

Wind Velocity
Experiment 2

Results and Conclusions:

1. Did wind speeds vary with altitude? _____ How do you know?
   _____
   _____

2. Wind prospectors fly helium-filled weather balloons instead of kites. What advantage might this give? _____
   _____
   _____

3. Compare your hypothesis with your results. How accurate was your hypothesis? _____
   _____
   _____

4. The kite rises because wind pressure on the lower face of the kite pushes it upward. What do you think would be the greatest angle a kite might reach on your inclinometer? _____ How could you find out? _____
   _____
   _____
   _____
   _____

5. If you were setting up a windmill farm in your area, how would you determine the best place for the farm? _____
   _____
   _____
   _____
   _____

Science Challenge: Set up an experiment to test this question: **Where is the best location for a windmill at your school?** Write your question, hypothesis, procedure, and materials list on another sheet of paper. Then test the hypothesis and record your conclusions.

EXPERIMENT 2

Creative Teaching Press/Energy

Wind Energy Applications

Windmills were used in Persia (now Iran) about 1000 A.D. Some researchers believe windmills were used even in the seventh century. Windmills are still used today to pump water and grind grain in many countries. Sailplanes, hang gliders, sailboats, and iceboats also depend on the wind and air currents for their energy. Most modern wind machines are used to turn **generators** to make electricity. Unfortunately, wind-generated electricity meets only one percent of the world's power needs.

Sailing with the Wind
(Math)

An airfoil surface maximizes air force. Sails, windmill blades, and wings use wind and airfoil shapes to create lift. As wind moves over the curved forward or upper surface of the airfoil, the molecules of air move more rapidly over the upper (or windward) surface relative to the lower (or leeward) surface and a partial vacuum results. The sail, or wing, is pulled upward (or forward). In sailing terms this is called lee suction.

Wind speed at sea is often measured in knots (nautical miles per hour). One knot is one and one-eighth miles per hour (1.125). Have students multiply the velocities on the Beaufort wind scale on page 21 by 1.125 to convert them to knots.

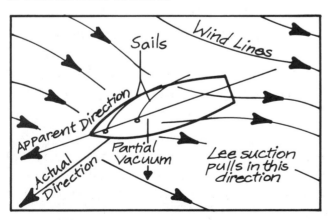

Moving Air
(Science Experiment/Oral Language)

Windmill blades are cut in the same manner as sails and airplane wings. They have a camber, or curve, that helps pull them forward like a sailboat. The surface that faces the wind is forced aside in the same manner as a kite is pushed into the sky. The pressure against the blade causes it to move.

Place a Ping Pong™ ball on the nozzle of a hair dryer. Turn the dryer on high. The ball stays suspended in the air directly above the nozzle. Invite students to explain their observations. (The airflow simultaneously reduces the air pressure on all surfaces of the ball. The curved surface of the ball can be compared to the curved surface of a blade. If the ball were flat on one side, it would move out of the air current in the direction of the curved side. Because the ball is curved on all surfaces, it is always pulled back to the center of the airflow.)

Extension Activities

Run Like the Wind
(Math)

The sun warms air unevenly. Warm air has low pressure and rises. Cool air has high pressure and moves downward. Winds blow from a region of high pressure to a region of low pressure. The difference in air pressure between regions determines wind velocity. The greater the difference in air pressure, the faster the wind velocity. Altitude also affects wind velocity. Eleven kilometers above the earth's surface, wind can move up to 135 meters per second (300 miles per hour). These fierce winds, called jet streams, blow in narrow, river-like bands from west to east.

An anemometer measures wind velocity. Velocity is recorded as meters per second. Calculate wind speed using the equation $r = d \div t$ (rate equals distance divided by time). Have students use the formula to measure each other's speed as they simulate wind. Time students with a stopwatch as they run fifty meters. Tell students their times and have them compute their velocity. Determine the average "wind" velocity of the class.

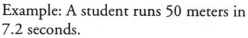

Example: A student runs 50 meters in 7.2 seconds.

$$50 \text{ m} = r \times 7.2 \text{ seconds}$$
$$r = 50 \div 7.2$$
$$r = 6.94 \text{ m per second}$$

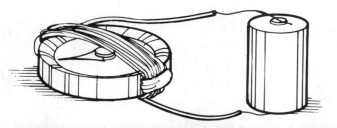

Wind Power
(History/Creative Writing)

Have students create a time line of the history of wind power; compare blades of windmills with blades of water turbines, listing similarities and differences; and/or design and illustrate new and fanciful uses for windmills (e.g., powering a lawn mower or portable radio).

Galvanometer
(History/Creative Writing)

A galvanometer detects small amounts of electric current. To make a galvanometer, give students a wire stripper, a simple compass, vinyl tape, and 120 cm of bell wire from the local hardware store. Have them follow these steps:

1. Strip 3 cm of insulation from one end of the wire. Measure the wire to a length of 12 cm.
2. Wrap the length of wire eight times around the compass. Wrap the wire evenly in one direction.
3. Strip 3 cm from the other end of the wire.
4. Touch each end of the wire to one end of a battery.

Have students describe which direction the needle moves when the galvanometer comes into contact with the battery. Students can use the galvanometer to test the zinc and penny battery constructed on page 22.

Water Energy

*What is harder than rock,
or softer than water?
Yet soft water
Hollows out hard rock.*

—Ovid, *Ars Amatoria*

Key Concepts of Water Energy

Water Energy—
safe, clean, non-polluting source of energy derived from moving or falling water

Water Energy Storage—
potential water energy stored behind dams in reservoirs

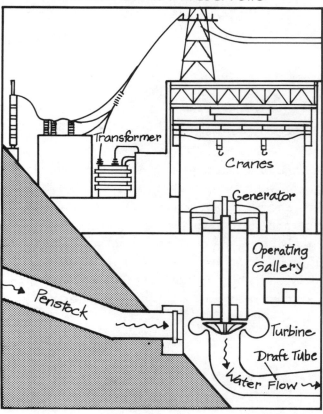

Water Energy Applications—
electricity for home and industrial use, and mechanical power for grinding grain

Water Energy

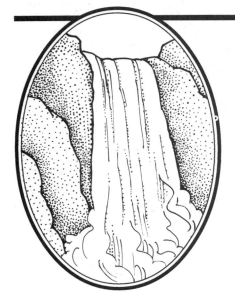

Wooden waterwheels have been used to drive machinery in flour mills and factories for thousands of years. Steam engines replaced wooden waterwheels in the nineteenth century. Today, water moving from a higher level to a lower level activates turbines that drive electric generators. This energy is called **hydroelectric power.** Hydroelectric power sources include dams, waterfalls, wave action, and tidal currents. **Hydraulic machines** use moving water as a source of power. The equipment used to generate hydroelectric power is expensive and involves extensive engineering and materials.

I'm Steamed
(Science Experiment)

Thermal or heat energy makes things move. Have students fill a teakettle about three-fourths full of water and place it on a heat source such as a stove or burner. Turn on the heat and wait until the kettle whistles. When the steam begins to come out, hold a pinwheel in the path of the steam and observe. The heat energy from the water changes to kinetic energy.

> **Caution:** Steam causes severe burns. Do not hold your arm, hand, or pinwheel too near the rising steam.

Hydraulic Turbines
(Social Studies)

There are two types of hydraulic turbines—reaction and impulse. James B. Francis of Lowell, Massachusetts developed the reaction turbine in 1849. The reaction turbine is most efficient when the head, the difference in height between the highest and lowest level of water, is less than 100 feet. Lester Pelton's wheel, an impulse turbine, works most efficiently when the head is at least 800 feet.

In a pumped-storage plant, generators run electric motors that pump water upward to a high-level reservoir during periods of low electrical demand by using the excess electricity available. During periods of high electrical demand, the facility uses water that flows down from the reservoir to produce electricity. Have students determine possible periods of high and low electrical demand that could occur in their geographical area. As a class, list the possibilities and rank them in order of highest to lowest probability.

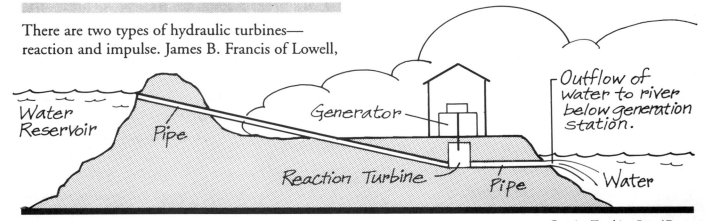

Water Energy Storage

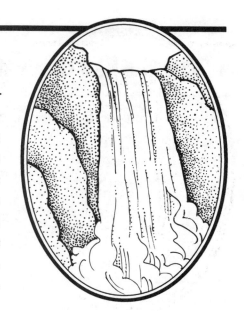

Dams are barriers that hold back water and often form reservoirs or lakes. Water stored behind dams holds potential energy. Hoover Dam in Nevada and Arizona was built between 1930 and 1935. Two years worth of the Colorado River's flow is stored behind the dam. The dam is 45 feet thick at the top and 660 feet thick at the base. It is as tall as a 60-story building. Before the dam could be built, geological and topographical studies were made. Modern dams, such as the Hoover, provide water for irrigation, assistance in flood control, improvement in the navigability of waterways, hydroelectric power, and recreational areas.

Dams of the World
(Model Building/Geography)

Have cooperative groups research one of the major dams of the world, such as:

Aswan Dam, Egypt
Daniel Johnson Dam, Canada
Guri Dam, Venezuela
Hoover Dam, Nevada and Arizona, U.S.A.
Itiapu Dam, Brazil and Paraguay
Kariba Dam, Zambia and Zimbabwe

Challenge students to locate their dam on a world map, construct a model, and give an oral presentation to the class. Encourage them to include the geography, history, construction, energy storage potential, and present day uses of the dam.

Pressure Pipe
(Math/Writing)

Use the pressure pipe from page 32 as a dam. Using the following formula, have students determine the volume of water in the pipe at each depth and fill in the chart. (Answers in the second column are provided for your convenience.) The formula for volume in cubic centimeters is *Volume = $\pi r^2 h$*. The pipe is 5 cm across, so the radius equals 2.5 cm (half the diameter). π(pi) = 3.14 and height (h) is in centimeters. The word formula reads: *Volume equals pi times radius squared times height.*
For a 30-cm water height, the equation would be:
$3.14 \times 2.5^2 \times 30 = 588.75$ cubic centimeters.

Height in centimeters	Volume in centimeters
30 cm	588.75
60 cm	1177.50
90 cm	1766.25
120 cm	2355.00

After students have made the calculations and completed the Water Pressure experiment on pages 33 and 34, have them respond to these questions: *What did you learn from this activity? Why is water pressure important in a dam?*

Water Pressure
(Science Experiment)

In Experiment #3 on pages 33 and 34, students observe how increasing water level increases water pressure. The pressure pipe is a simulation of a dam. The crucial issue in water pressure is water height, not water volume. Prepare for the experiment by having parent volunteers help build pressure pipes as described on page 32.

Creative Teaching Press/*Energy*

Pressure Pipe

This pressure pipe can be used with the experiments and activities on pages 31–35.

Materials
- 5' plastic sprinkler pipe (2" diameter)
- 2" end cap
- sprinkler pipe adhesive
- measuring tape
- drill with 1/8" bit
- small rat tail file (optional)
- scissors
- masking tape
- safety goggles
- safety gloves

SAFETY PRECAUTION: Adults should build these pipes for student use in Experiment #3.

Procedure:

1. Coat 1" of one end of the 5' pipe with sprinkler pipe adhesive. Twist the end cap onto the coated pipe, sealing it against water leaks.

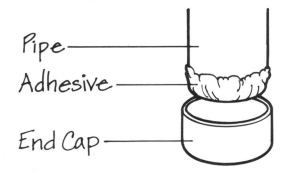

2. Drill a hole in the side of the pipe, just above the cap. Drill holes at 30-cm increments from this hole to the top of the pipe. Be sure the holes are aligned.

Drilling Tip: Cover spots to be drilled with a piece of masking tape. Use scissors to mark where to drill on the tape. Tape will help keep the drill bit from slipping.

3. Clean the holes with a rat tail file or by plunging the drill bit in and out of each hole. This will smooth the openings of the holes, allowing water to flow more smoothly.

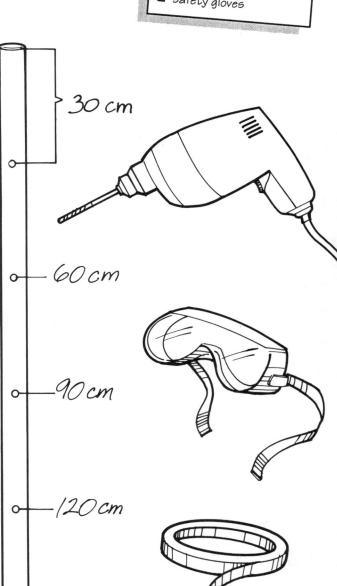

Name:

Water Pressure
Experiment 3

EXPERIMENT 3

Question: How does increased water pressure affect water flow?

Hypothesis: _____
_____
_____

Materials
- ☐ pressure pipe (page 32)
- ☐ modeling clay
- ☐ garden hose
- ☐ water supply
- ☐ measuring tape
- ☐ chart paper

Procedure:

Step 1

Plug all the holes in the pressure pipe, except the top hole, with large lumps of clay.

Step 2

Place the pipe on an elevated surface such as an outdoor table or step so water squirts freely from the hole.

Step 3

Hold the pipe perpendicular to the ground while the hose fills the pipe with water. Maintain the level so water continually fills the pipe but does not bubble over the top of your "dam."

Step 4

Measure and chart the distance water is ejected from the hole.

Step 5

Keeping the water flow from the hose constant, unplug the holes, one at a time. As you unplug each hole, measure the distance water is ejected. Chart the results.

Level of Hole	Distance Water Ejected
30 cm	
60 cm	
90 cm	

Creative Teaching Press/*Energy*

Name:

Water Pressure
Experiment 3

Results and Conclusions:

1. Describe any variation in the strength or distance water ejected when comparing the 30-cm and 90-cm holes. _____
_____
_____

2. Did your results form a mathematical pattern in your chart? _____ If so, describe the pattern. _____
_____
_____

3. Which hole emitted the strongest flow of water? _____
The weakest flow of water? _____

4. Graph the information from the two-column chart on page 33. Use the first column as the *x axis* and the second as the *y axis*. What distances would you predict water would eject from holes drilled at 105 cm, 130 cm, and 165 cm? _____
_____
_____

5. How did the results compare with your hypothesis? _____
_____
_____

6. How does what you learned about water pressure apply to how a hydroelectric dam works? _____
_____
_____
_____

Science Challenge: Set up an experiment to test this question: How does the size of the opening in a water pipe affect water flow? Write your question, hypothesis, procedure, and materials list on another sheet of paper. Then test the hypothesis and record your conclusions.

Water Energy Storage

The Pelton Wheel
(Model Building)

As many as four jets of water strike the buckets in the Pelton wheel, causing continuous and efficient conversion of potential energy to kinetic energy. Invite a group of students and adults to build several Pelton wheels to be used in the next activity. Each group will need aluminum cans, gloves, utility shears, a stapler, an awl, a small ball of clay, and a metal hanger. Have groups follow these directions:

Caution: Be careful of sharp edges.

1. Cut 1" off the top of an aluminum can.
2. Cut straight from the top opening to within 1" of the lower shoulder of the can.
3. Cut along the opposite side in the same manner.
4. Carefully cut the two "halves" in half to make four "blades."
5. Cut the four blades in half to make eight blades.
6. Fold each blade at a 45° angle. Staple them on the fold.
7. Crease the length of each blade in the opposite direction of the natural curve.
8. Punch a hole in the center of the bottom of the can with the awl. Punch a hole in the bottom of another aluminum can that has been cut to a 1" height.
9. Join the two pieces by placing one inside the other to complete the Pelton Wheel.
10. Place a straightened piece of metal hanger through the holes in the Pelton wheel to hold during the Moving Wheels activity that follows.

Moving Wheels
(Science Experiment)

Invite cooperative groups to use the aluminum Pelton wheels from the previous activity and the pressure pipes from page 32 to discover how a hydraulic impulse wheel works. The water stream should hit the blades of the wheel. Use one pipe with the water source coming from a hose. If possible, obtain a second water source (e.g., a large pitcher) to try to move the wheel with more than one water jet. Have students predict which holes provide water streams that will turn the wheels. Encourage groups to experiment with the number of water jets they can position to simultaneously move more than one wheel.

Water Energy Applications

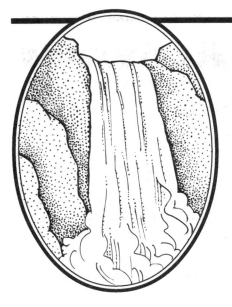

Waterwheels were invented about 2000 B.C. and probably originated in Greece. One of the first uses of dams was to provide water for field **irrigation.** One of the first uses of the steam engine was to drive a pump that raised water into a millpond. The millpond's spillway drove a waterwheel. Today, dams serve multiple purposes such as water storage, recreation, and generation of **hydroelectric power.** Dams and waterwheels have seen many changes and refinements over the last two centuries.

Tide-Powered Generators
(Research/Debate)

Tide-powered generators have reaction turbines placed in the way of changing tides. Harbors experiencing extreme differences in high and low tides benefit most from these turbines. The energy harnessed can produce electricity.
Have students research which locations in the world have the greatest tidal ranges. (For example, the Bay of Fundy in Canada experiences a tidal range of 47 feet.) Then have students research what type of impact tide-powered generators might have on marine life. Invite students to debate these topics: 1) The benefits of installing tide-powered generators outweigh negative environmental impacts. 2) The cost of installing tide-powered generators is justified.

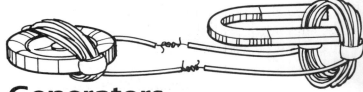

Generators
(Science Experiment)

Generators convert water energy into electricity, giving us hydroelectric power. Students can make their own generators by following these steps:

1. Strip 3 cm of covering from the ends of 60 cm of wire.
2. Make ten loops of wire 5 cm in diameter. Secure the loops with electrical tape.
3. Connect the two wires to the wires of a galvanometer (see page 27).
4. Push a magnet in and out of the coils.
5. Observe the galvanometer. It should record electrical activity.

Generators rotate wire coils in a "tube" of magnets or rotate magnets inside coils to create electricity. (The energy of the falling water from a dam causes the actual physical rotation.) A generator is basically a motor shaft being turned to send electricity back out of the wires. Have students experiment with different-sized magnets and different numbers of loops to see which generates the most activity.

Extension Activities

Waterwheels
(Model Building)

Have students build waterwheels with paper cups, masking tape, straws, straight pins, and pencils. Ask them to follow these steps:

1. Tape two straws perpendicular to one another.

2. Securely tape paper cups underneath the end of each straw.

3. Press a straight pin through the point where the two straws cross and press the pin into the eraser of a pencil.

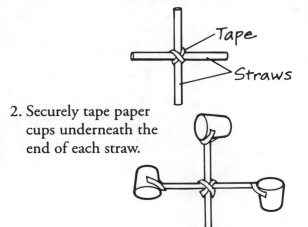

4. Place the wheel underneath a water source such as a drinking fountain. Observe.

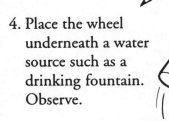

Challenge students to use several waterwheels to move water from one location to another.

Water Cycle I
(Science Experiment)

The water cycle moves water between the earth's surface and the atmosphere. In this activity, students observe the basic processes that form the water cycle—condensation, evaporation, and precipitation.

1. Fill two same-sized jars three-fourths full of water.
2. Securely cover one jar with a lid.
3. Place both jars by a sunny window for several days.
4. Record your observations in your journal.

Water Cycle II
(Demonstration)

Fill a large saucepan about one-fourth full of water. Place the pan on a hot plate or stove. Bring the water to a boil. Fill a small saucepan with ice cubes. Put on a protective mitt and hold the small pan over the boiling water of the large pan. Watch what happens to the bottom of the small pan. Record your observations in your journal.

Fables
(Literature/Writing)

Read selections from *Aesop's Fables* as retold by Ann McGovern. Have students write and illustrate a fable with a moral, using sun, water, or wind energy as a theme. Compile student fables and make them available for checkout.

Fossil Fuels

*Bright-flaming, heat-full fire,
The source of motion.*

—Guillaume Du Bartas,
Devine Weekes and Workes

Key Concepts of Fossil Fuels

Fossil Fuels—
fossilized plant and animal matter found as coal, crude oil, and natural gas

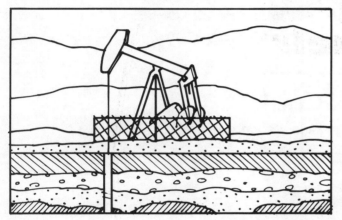

Fossil Fuel Storage—
maintenance of mined or pumped oil, coal, and natural gas, refined into forms (fractions) and stored in tanks

Fossil Fuel Applications—
electricity, transportation, heating, and cooking

Kerosene- Jet Fuel and Lighting

Lubricating Oil

Fuel Gas

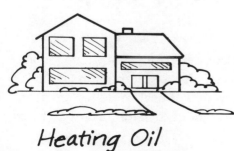

Heating Oil

Gasoline

Fossil Fuels

Fossil fuels are made from the remains of plants and animals exposed to extreme heat and pressure over extended periods of time. **Coal, crude oil,** and **natural gas,** removed from underground deposits or wells, are the primary fossil fuels. Coal has served as a heating fuel for thousands of years. Crude oil is **refined,** or **fractioned,** into many useful compounds such as gasoline, heating oils, and lubricants. Natural gas is used as an industrial and domestic fuel for heating and cooking. Although present reserves of fossil fuels are considerable, supplies are limited and **nonrenewable.** Eventually fossil fuels will have to be replaced with other energy sources.

Gasoline
(Critical Thinking/Writing)

In the 1850s, when gasoline was first separated from oil, people discarded the gasoline because they did not understand how to use it. Combustion engines now require gasoline or diesel fuel. Gasoline is an explosive compound. Have cooperative groups write safety codes for handling gasoline.

Kerosene
(Research/Art)

Kerosene forms jet fuel when combined with alkylates. Kerosene burns, but does not explode. Discuss with the class why this might be an important consideration for flight use. Have students research kerosene, safety precautions, and present-day applications and create a cartoon strip to share their findings.

Coal
(Science Demonstration/Writing)

Coal supplies half the electrical energy for the United States and about two-thirds of the energy for the rest of the world. Have students examine a piece of coal under a magnifying glass and describe their observations. Does the coal have luster, layers, or fossils? Fossils are evidence of past plant or animal life.

Put the coal in a metal container. Sprinkle lighter fluid over it and light it. Have students feel the heat energy given off by the reaction, observe the black soot pollution as a waste product, and record observations in science journals.

Adult supervision required.

Dwindling Supplies
(Debate/Ecology)

At our present rate of consumption, we will exhaust our oil and gas supplies in about 100 years. The supply of coal could last the next 350 years. Have students debate this topic: Fossil fuel use should be made illegal worldwide.

Fossil Fuel Storage

Crude oil, or **petroleum,** is pumped from wells and piped into storage bins where it awaits refining. Crude oil is refined by **fractional distillation,** producing gasoline, kerosene, diesel oil, fuel oil, lubricating oil, and asphalt. These products are stored in large tanks and transported by trucks, pipes, and oil tankers. Natural gas is transported by large pipelines or, as a liquid, in refrigerated tankers. Natural gas is stored in large tanks.

Gas Storage Tanks
(Writing/Art)

Rows of tanks, or tank farms, store oil in different forms or fractions. Many pipes run from the refinery to the tanks and from tank to tank. A high earthen dike, or berm, surrounds tank fields. A fence topped with barbed wire surrounds the dike. Have students illustrate and write a paragraph speculating the reasons for triple safety measures. (Earthen dikes help contain oil in case of leakage. The fence and barbed wire help prevent vandalism.)

Carbonization
(Science Experiment)

pages 42-43

Fuel can be stored in carbon form. All living things contain carbon. The human body is 18% carbon by weight. Coal, a fossil fuel of plant origin, is almost entirely composed of carbon. Coal is formed when plant remains are exposed to extreme pressure and heat over an extended period of time. In Experiment #4 on pages 42 and 43, students observe the effects of heat on several substances. The substances will be reduced to their carbon content, a process called *carbonization*.

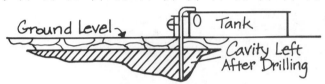

Natural Gas Storage
(Model Building/Writing)

Natural gas is stored in large underground cavities left behind when oil is pumped out. Supplies are stored in summer and fall and used in cold weather when demand is greater.

In the past, gas was stored in vertically-expanding tanks. The tanks were set in frames so, as more gas was stored, the tank could expand to meet the increased volume. Have students build a model of either type of natural gas storage. Students can include a paragraph listing the advantages and disadvantages of the design they chose to build. Share with students that engineers build models to determine value, safety, and construction costs.

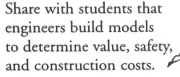

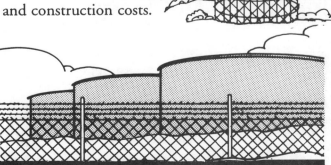

Name:

Carbonization
Experiment 4

EXPERIMENT 4

Question: How can organic matter be reduced to carbon?

Hypothesis: _____
_____
_____

Materials
- ☐ sugar, flour, and salt (1/2 cup each)
- ☐ spoon
- ☐ saucepan
- ☐ stove or hot plate
- ☐ aluminum foil
- ☐ balance scale

Procedure:

Step 1
Measure and record the weight of the sugar.

Step 2
Place the sugar in the pan. Place the pan on a hot plate and slowly heat the sugar. Record the time needed to convert the sugar to its black carbon form on the chart to the right.

Step 3
Save the carbonized sugar on a piece of foil. Weigh it. How much weight was gained or lost?

Be sure foil is on a plate to keep from scorching desks or countertops.

Step 4
Wash the pan. Then repeat steps 1–3 for the flour and salt.

	Sugar	Flour	Salt
Weight Before			
Weight After			
Difference			
Time Needed for Carbon-ization			

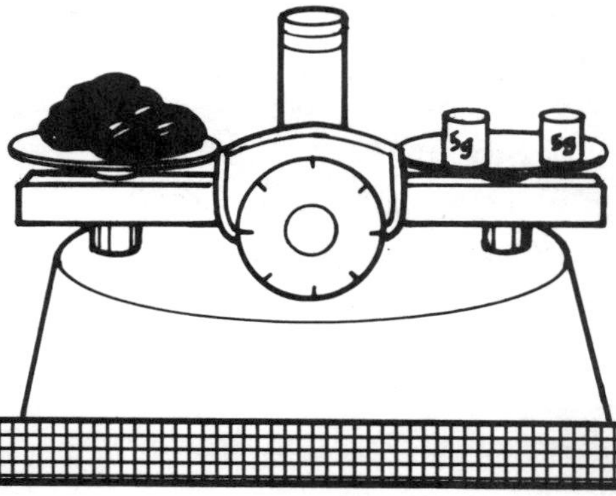

42   Creative Teaching Press/*Energy*

Name:

Carbonization
Experiment 4

EXPERIMENT 4

Results and Conclusions:

1. How were the three substances similar before and after the experiment? _____
_____
_____

2. Which of the three contained the least carbon? How do you know?
_____
_____
_____

3. Speculate about other kitchen supplies. What else could you carbonize? _____
_____
_____

4. Coal and oil were created from plant and animal matter under pressure over a long period of time. From what you have learned, what else might cause carbonization? _____
_____
_____
_____

5. What other questions do you have about carbonization? _____
_____
_____
_____
_____
_____

Science Challenge: Set up an experiment to test this question: **How are coal and sugar related?**

Write your question, hypothesis, procedure, and materials list on another sheet of paper. Then test the hypothesis and record your conclusions.

Creative Teaching Press/*Energy*

Fossil Fuel Storage

Oil Wells
(Geology/Art)

Crude oil is found in five different geological structures—seepages, anticlines, stratigraphic traps, faults, and salt domes. Have students design and illustrate their own oil wells. Ask them to show their oil wells drilling into one of the five types of geological structures. Invite students to share their models with the class.

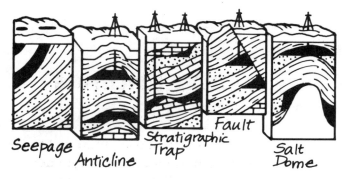
Seepage, Anticline, Stratigraphic Trap, Fault, Salt Dome

Removing Oil
(Science Experiment)

Students investigate the most efficient way to use pumps to remove oil from reservoirs. Students make oil pumps by using clear plastic bottles with spray-pump attachments. Plastic tubing should extend from the pump attachment to the bottom of each container. Invite students to follow these steps:

1. Place 1–2 cups of clean, small pebbles in the bottle.
2. Pour 100 mL of corn oil into each bottle.
3. Carefully start pumping as much oil as possible out of the container into a 100-mL graduated cylinder.
4. Measure and record.
5. Clean the bottle and cylinder and repeat the steps using the following combinations: 50 mL cold water/50 mL oil; 50 mL hot water/50 mL oil/10 drops of detergent.

Caution: Wear safety goggles. Collect used oil in a separate bottle.

Refining Oil
(Science Experiment)

Petroleum products such as gasoline, kerosene, diesel oil, and lubrication oils are distilled or separated from crude oil in refineries. Oil pumped through tubes is boiled and then travels from bottom to top in a fractioning tower. Gasoline vaporizes first, since it has the lowest boiling point. Kerosene vaporizes second and diesel fuel third. Gasoline is drawn off at the top of the tower. Lubricating oils are distilled at the bottom. The following steps illustrate the distillation process:

1. Dissolve one-quarter cup salt in two cups of water.
2. Boil it in a saucepan with a too-large lid hanging over the edge. Place a cup under the lid to catch the distilled water droplets as shown.

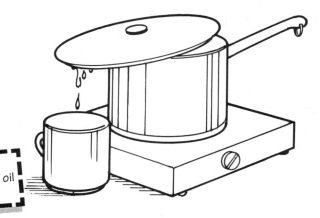

3. Allow the boiled water to cool.
4. Compare the taste of the salty water in the pan with the distilled water in the cup.

Have students respond to the following questions in their science journals: *What do you notice about the difference in tastes? Does the distilled water taste less salty? What did this experiment demonstrate? How is distillation like fractioning?*

Fossil Fuel Applications

Gasoline and fuel oil run **internal combustion engines** which turn generators to create electricity. Ships, trains, trucks, and automobiles burn fuel in internal combustion engines or heat water to make steam to run **turbines.** Coal and natural gas are burned to create steam to run turbines and to heat homes and workplaces. Kerosene is a lighter fuel than gasoline and is used as jet fuel. Engines that burn fuel in a separate fire box outside the cylinder, such as those used on locomotives, are called **external combustion engines.**

Internal Combustion Engines
(Art)

An internal combustion engine is one in which fuel combustion takes place in a confined space, producing expanding gases used to provide mechanical power. Most internal combustion engines use a piston-type gasoline engine. The piston attaches to a connecting rod and crankshaft, allowing the piston an up-and-down motion. The downward motion of the piston draws in a mixture of air and gasoline. The upward motion of the piston compresses the air-gasoline mixture into a small space. A spark plug ignites the air-gasoline mixture which forces the piston downward again. This downward motion, or power stroke, turns the crankshaft which drives the wheels of the car.

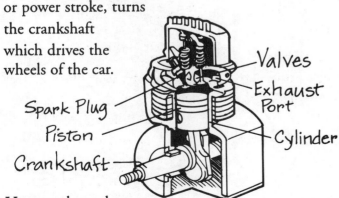

Have students draw and label an internal combustion engine.

External Combustion Engines
(Art/Model Building)

Steam engines, such as locomotives, burn fuel to create steam. The steam is forced into a cylinder, driving the piston down. This motion pushes a long crank or arm which turns the wheels. Since the heat is created outside the cylinder containing the piston, it is called an external combustion engine.

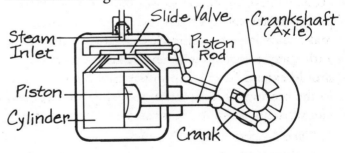

Place a bicycle upside down on a table and push the pedals around. This turns the crank and wheel similar to the way a steam engine drives a locomotive. Invite a train engineer or collector of model trains to come to class and share his or her knowledge. Students will also enjoy books and videos about the development of locomotives. Have students draw an external combustion engine and label the parts or build a model of a steam engine or train.

Creative Teaching Press/*Energy*

Extension Activities

Carpooling
(Ecology/Math)

Carpooling saves valuable fossil fuels and money. Have students write carpooling situations on index cards and challenge classmates to solve them.

> If four neighbors each drive daily roundtrips of 25 miles, how much fuel and money would they save if they carpooled? Choose the type of cars and assume 48 five-day work weeks.

Drake's Folly
(Research/Oral Language)

Edwin L. Drake (1819–1880), a retired train conductor, was interested in the oil and tar deposits around Titusville, Pennsylvania. Drake hoped to separate kerosene from petroleum, first done in Canada by Abraham Gesner (1797–1864). Drake believed kerosene could replace whale oil for home lighting. Whales had been hunted to scarcity, and their oil was very expensive. Drake used a steam-operated drill. After three months of drilling, he hit oil. The oil sold for $20 a barrel. Others copied Drake's work, since he did not patent the process. Within three years, the cost of a barrel of oil dropped to 10¢.

Have students work in cooperative groups to research a topic of interest and prepare a presentation. Topics might include the whaling industry and uses of whale oil, the oil industry, the patenting process, oil drilling, oil prices, and so on.

Electric Meters
(Ecology/Journal Writing)

Have students find their electric meter at home and record their daily electrical use for one week. Have them calculate the total number of kilowatt hours used for that week. Also ask them to list ways their families can conserve electrical energy, and discuss these ideas at home.

Crazy Putty
(Science Experiment)

Plastics are made of long chains of carbon-based molecules. When many of these chains join together, it is called a polymer. Have students make their own polymer by following these steps:

1. Place two or three ounces of white glue in a paper cup. (Add food coloring, if desired.)
2. Add an ounce or two of sodium borate solution (1 quart of water to one cup of borax detergent).
3. Stir and remove the solid from the cup.
4. Push, pull, roll, drop, and stretch the "crazy" putty. Store the material in a closed, plastic bag.

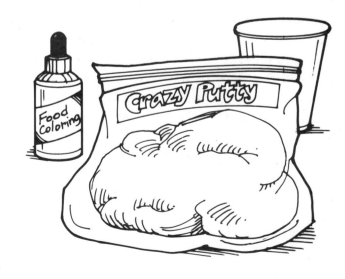

Extension Activities

OPEC
(Oral Language/History)

Have students interview family and friends about the Organization of Petroleum Exporting Countries (OPEC). Have them ask these questions: *When was the organization formed? What nations belong to OPEC? What is the purpose of OPEC? Why did OPEC place an oil embargo on the United States in 1973? What was the result of the oil embargo?* Encourage students to verify information with written sources before reporting findings to the class. Have students present information in a creative format such as a television newsbrief, a visual time line, or a series of captioned overhead illustrations.

Heat Transfer
(Science Demonstration)

Fossil fuels burn to produce heat energy. Heat energy can be transferred by conduction, convection, and radiation and used to do work.

1. *Conduction* occurs when hot, rapidly vibrating molecules bump into cooler, more slowly vibrating molecules. Heat energy is transferred and the cool object becomes hot. Soldering irons and stoves use heat conduction.
2. *Convection* occurs when a mass of heated gas or liquid rises to a cooler area. Space-heating systems use heat convection.
3. *Infrared (heat) radiation* consists of electromagnetic waves that move through matter or a vacuum. The heat we feel from glowing charcoal or a lightbulb is from infrared radiation.

Gather two candles, a knife, a six-foot length of wire, matches, and paper towels. Demonstrate heat transfer to students by following these steps:

1. Wrap the wire around two metal supports (e.g., two upside-down desk or chair legs).
2. Carve wax shavings from one of the candles.
3. Place the wax shavings along the length of the wire.
4. Place paper towels beneath the wire.
5. Light the other candle and hold it near one end of the wire.
6. Ask students to observe the wax shavings.
7. Have students record their observations.

The wax shavings will melt and drip in order, beginning with the shaving nearest the candle flame. Ask students how this observation illustrates heat transfer. (This is an example of conduction.) Invite cooperative groups to design their own experiments to demonstrate the three types of heat transfer.

Nuclear Energy

*The unleashed power of the atom
has changed everything
save our modes of thinking,
we thus drift toward unparalleled catastrophes.*

—Albert Einstein,
Letter to William Fauenglass, May 16, 1953

Key Concepts of Nuclear Energy

Nuclear Energy—
energy released during nuclear reactions such as fission or fusion

This symbol means radiation source.

Nuclear Reactors—
power plants in which the controlled release of nuclear energy takes place

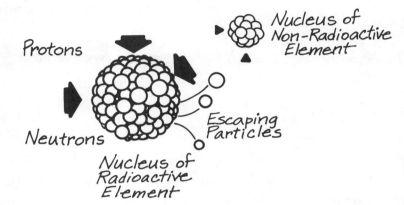

Nuclear Energy Applications—
electrical power to run nuclear submarines or provide energy for households

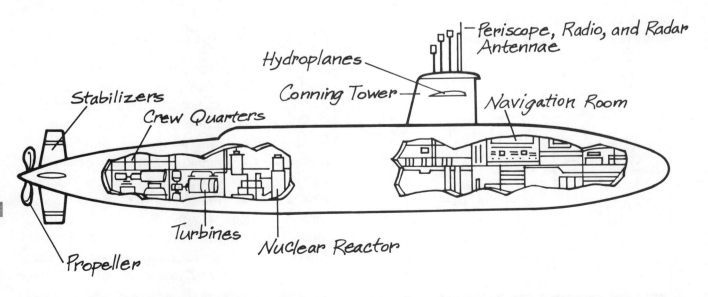

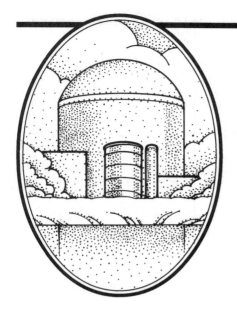

Nuclear Energy

Nuclear energy is obtained from nuclear reactions. Nuclear reactions involve changes in the nucleus of an atom. **Nuclei** are the small, dense centers of atoms composed of protons and neutrons. Some elements, such as uranium, plutonium, and hydrogen, have greater nuclear energy potential than other elements. Releasing nuclear energy depends on atomic structure and is difficult to accomplish. Nuclear energy can be released through two processes—fission and fusion. **Fission** is the breaking up of large nuclei into smaller ones. **Fusion** is the building up of small nuclei into larger ones.

Fission
(Research)

Fission is a nuclear reaction in which a heavy atomic nucleus, such as uranium, spontaneously splits into two parts upon capturing a free neutron. A heavy atomic nucleus has more than 26 protons, the number of protons in an iron atom. When the nucleus splits, it releases two or three neutrons and a large quantity of energy. The released neutrons hit other nuclei, creating a chain reaction. Have students list three elements that would release energy if forced to undergo fission. (See Periodic Table of Elements, page 54. Possible answers include any element with an atomic number greater than iron.)

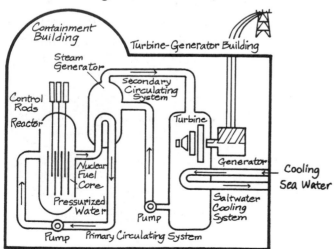

Fusion
(Research)

Atoms with a simpler atomic structure than iron (lower atomic number) release energy when their nucleus is fused or joined with another nucleus. Have students use the Periodic Table on page 54 to list five elements which might undergo fusion. (Possible answers include any element with an atomic number lower than iron.)

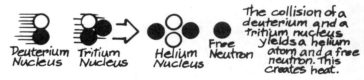

The collision of a deuterium and a tritium nucleus yields a helium atom and a free neutron. This creates heat.

Radioactive Film
(Science Experiment)

Radiation was discovered when a radioactive source was placed near a photographic plate. Some workers wear photographic film badges for protection from radiation exposure. Obtain a badge or undeveloped film and a weak radiation source, such as pitchblende (an ore containing some radioactive isotopes of uranium and radium) or an old smoke detector containing americanium. Place the badge or film near the radiation source for ten minutes or longer. Develop the film, and observe the results.

Nuclear Reactors

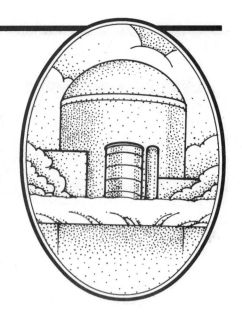

The controlled release of nuclear energy takes place in **nuclear reactors.** Nuclear reactors currently only use nuclear **fission.** They make use of the elements **uranium** and **plutonium** which are obtained from mineral ores rich in these elements. The ores are first treated by a process called **enrichment** before being put into the reactors. Energy released by fission is used to power electrical generators. The nuclear process is controlled, and only the desired amount of heat is released.

Chain Reaction
(Demonstration)

Uranium is an unstable element because nuclei break apart more easily than most elements. When the nucleus splits, it releases energy and neutrons. The neutrons strike other nuclei which also split, releasing more energy and neutrons. Help students visualize this chain reaction by having cooperative groups set dominoes on end in a bowling pin pattern. Have them knock over the first domino and watch the chain reaction.

Nuclear Fission
(Demonstration/Science Experiment)

Experiment #5 on pages 52 and 53 is a demonstration only; however, students should still fill out experiment worksheets. Students will observe a simulation of a nuclear fission chain reaction. This simulation is quick and dramatic. The fire cloud simulates the chain reaction occurring in nuclear fission.

Half-Life Toss
(Demonstration)

Elements used as fuel in nuclear reactors last about two years. When the elements can no longer be used, they are *spent*, becoming nuclear waste. Because the element continues to break down, it is considered unstable. The time it takes for half the atoms of an unstable radioactive element to "run down," or become stable, is the element's *half-life*. The half-life of radioactive radium is 30 seconds, while the half-life of radioactive uranium is 4,510,000,000 years! Follow these steps to demonstrate half-life:

1. Have each student construct a data table with two columns titled *Half-life* (tosses) and *Atoms Remaining* (heads up).
2. Give each student a penny. On the first line of the data table, students record the beginning data (0 under *Half-life* and the number of students under *Atoms Remaining*).
3. Have all students stand and toss their coins at once. This represents one half-life. All coins landing as tails are "decayed." Students whose pennies have decayed stay seated.
4. Count the students still standing and record the appropriate numbers in the data table.
5. Continue this procedure until no one is left standing.

Creative Teaching Press/*Energy*

Name: _____

Nuclear Fission
Experiment 5

EXPERIMENT 5

Question: How can the chain reaction of nuclear fission be simulated?

Hypothesis: _____
_____
_____

Materials
- ☐ cup of flour
- ☐ Bunsen burner or candle
- ☐ straw
- ☐ safety goggles
- ☐ baseball cap

Procedure:

> WARNING: Teacher demonstration only. Students fill out experiment worksheets while making observations.

Step 1
Place the burner or candle on a solid surface. For safety, clear a space of eight to ten feet around the candle.

Step 2
Wear your safety goggles and cap since you will be expecting a flash of energy. The cap will help protect your hair from flame. Light the burner or candle and darken the room.

Step 3
Place a quarter teaspoon of flour in the center of your palm.

Step 4
Hold your hand at the level of the candle top, but about two inches away from the flame.

Step 5
Blow sharply through the straw towards the flour at a 45° angle. The flour "cloud" will shoot up into the flame. Conduct several trials.

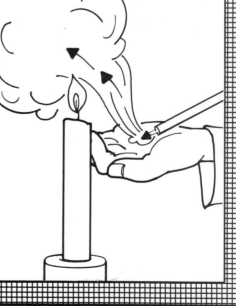

Creative Teaching Press/*Energy*

Name:
Nuclear Fission
Experiment 5

EXPERIMENT 5

Results and Conclusions:

1. Describe the reaction that occurred when flame and flour met.

2. Once fission begins in a reactor, the process causes nuclei that have been split to split other nuclei. The process rapidly repeats itself. How has your model demonstrated this rapid release of energy?

3. Why do you think it is important to control nuclear reactions?

4. What other household supplies might give a similar reaction—salt, baking powder, or baking soda?

Science Challenge: Set up an experiment to test this question: **What other way can a chain reaction be simulated?** Write your question, hypothesis, procedure, and materials list on another sheet of paper. Then test the hypothesis and record your conclusions.

Creative Teaching Press/*Energy*

Periodic Table of the Elements

Group	I-A	II-A	III-B	IV-B	V-B	VI-B	VII-B	VIII-B			I-B	II-B	III-A	IV-A	V-A	VI-A	VII-A	VIII-A
1	1 **H** Hydrogen 1.0																	2 **He** Helium 4.0
2	3 **Li** Lithium 6.9	4 **Be** Beryllium 9.0											5 **B** Boron 10.8	6 **C** Carbon 12.0	7 **N** Nitrogen 14.0	8 **O** Oxygen 16.0	9 **F** Fluorine 19.0	10 **Ne** Neon 20.2
3	11 **Na** Sodium 23.0	12 **Mg** Magnesium 24.3											13 **Al** Aluminum 27.0	14 **Si** Silicon 28.1	15 **P** Phosphorus 31.0	16 **S** Sulfur 32.1	17 **Cl** Chlorine 35.5	18 **Ar** Argon 40.0
4	19 **K** Potassium 39.1	20 **Ca** Calcium 40.1	21 **Sc** Scandium 45.0	22 **Ti** Titanium 47.9	23 **V** Vanadium 50.9	24 **Cr** Chromium 52.0	25 **Mn** Manganese 54.9	26 **Fe** Iron 55.9	27 **Co** Cobalt 58.9	28 **Ni** Nickel 58.7	29 **Cu** Copper 63.5	30 **Zn** Zinc 65.4	31 **Ga** Gallium 69.7	32 **Ge** Germanium 72.6	33 **As** Arsenic 74.9	34 **Se** Selenium 79.0	35 **Br** Bromine 79.9	36 **Kr** Krypton 83.8
5	37 **Rb** Rubidium 85.5	38 **Sr** Strontium 87.6	39 **Y** Yttrium 88.9	40 **Zr** Zirconium 91.2	41 **Nb** Niobium 92.9	42 **Mo** Molybdenum 95.9	43 **Tc** Technetium 98.0	44 **Ru** Ruthenium 101.0	45 **Rh** Rhodium 102.9	46 **Pd** Palladium 106.4	47 **Ag** Silver 107.9	48 **Cd** Cadmium 112.4	49 **In** Indium 114.8	50 **Sn** Tin 118.7	51 **Sb** Antimony 121.8	52 **Te** Tellurium 127.6	53 **I** Iodine 126.9	54 **Xe** Xenon 131.3
6	55 **Cs** Cesium 132.9	56 **Ba** Barium 137.4	57-71	72 **Hf** Hafnium 178.5	73 **Ta** Tantalum 181.0	74 **W** Tungsten 183.9	75 **Re** Rhenium 186.2	76 **Os** Osmium 190.2	77 **Ir** Iridium 192.2	78 **Pt** Platinum 195.1	79 **Au** Gold 197.0	80 **Hg** Mercury 200.6	81 **Tl** Thallium 204.4	82 **Pb** Lead 207.2	83 **Bi** Bismuth 209.0	84 **Po** Polonium 210.0	85 **At** Astatine 210.0	86 **Rn** Radon 222.0
7	87 **Fr** Francium 223.0	88 **Ra** Radium 226.0	89-103	104 **Db** Dubnium (261.0)	105 **Jl** Joliotium (262.0)	106 **Rf** Rutherfordium (263.0)	107 **Bh** Bohrium (262.0)	108 **Hn** Hahnium (265.0)	109 **Mt** Meitnerium (268.0)									

Atomic Number — Atomic Symbol — Element Name — Atomic Weight (rounded to the nearest tenth)

Elements 110-111 are not included in this periodic table, because their existence is still being disputed among scientists.

6	57 **La** Lanthanum 138.9	58 **Ce** Cerium 140.1	59 **Pr** Praseodymium 140.9	60 **Nd** Neodymium 144.2	61 **Pm** Promethium 147.0	62 **Sm** Samarium 150.4	63 **Eu** Europium 152.0	64 **Gd** Gadolinium 157.3	65 **Tb** Terbium 158.9	66 **Dy** Dysprosium 162.5	67 **Ho** Holmium 164.9	68 **Er** Erbium 167.3	69 **Tm** Thulium 168.9	70 **Yb** Ytterbium 173.0	71 **Lu** Lutetium 175
7	89 **Ac** Actinium 227.0	90 **Th** Thorium 232.0	91 **Pa** Protactinium 231.0	92 **U** Uranium 238.0	93 **Np** Neptunium 237.0	94 **Pu** Plutonium (244.0)	95 **Am** Americium (243.0)	96 **Cm** Curium (247.0)	97 **Bk** Berkelium (247.0)	98 **Cf** Californium (251.0)	99 **Es** Einsteinium (252.0)	100 **Fm** Fermium (257.0)	101 **Md** Mendelevium (258.0)	102 **No** Nobelium (259.0)	103 **Lr** Lawrencium (260.0)

Approximate values for radioactive elements are listed in parentheses.

Nuclear Energy Applications

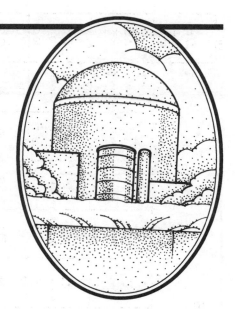

Nuclear fission is an efficient source of energy. Nuclear fission has been successfully used in land-based **nuclear power plants** and portable applications such as nuclear-powered submarines and other ships. There was speculation that nuclear fission would drive space craft by the 1970s, but that has not yet happened. Nuclear fission has not proved useful in operating electronic equipment onboard research satellites, but the radioactive decay process has been used as a source of heat.

Nuclear Vessels
(Oral Language/Writing)

The S.S. Savannah was the first steamship to cross the Atlantic Ocean (1819). In the early 1960s, another Savannah was built, this one powered by nuclear energy. The United States Navy is the largest consumer of nuclear power in the world. From submarines to aircraft carriers, much of the U.S. Navy uses nuclear power. Fuel elements (e.g., uranium) for a nuclear submarine last about four months. Nuclear fuel is used to create steam for turbines that generate electricity. The electricity runs electric motors that turn the submarine's propellers.

Have cooperative groups evaluate the advantages and disadvantages of nuclear-powered submarines and ships. Ask students if they would want to be stationed onboard a nuclear submarine, and write group responses. Invite groups to share their answers with the class. Challenge students to bring in books or articles about nuclear-powered submarines.

Turbines and Jet Engines
(Critical Thinking)

Nuclear power stations are built around a central fission reactor. This type of power station generates large amounts of electricity. Nuclear fission creates heat. Heat is combined with water to produce steam. Steam is used to run turbines that generate electricity. Turbines are rotary engines that use a continuous stream of steam or water to turn a shaft. The shaft drives an electric generator. Steam turbines, rather than water turbines, generate most of our electricity. Have students compare the steam turbine and jet engine on page 56. Have students chart similarities and differences between these turbines.

Steam Turbine	Jet Engine
Uses steam	Uses jet fuel

Name:

Turbines and Jet Engines

Directions: Compare the diagram of the nuclear-powered, steam-driven turbine with the diagram of the jet engine. Then answer the questions below.

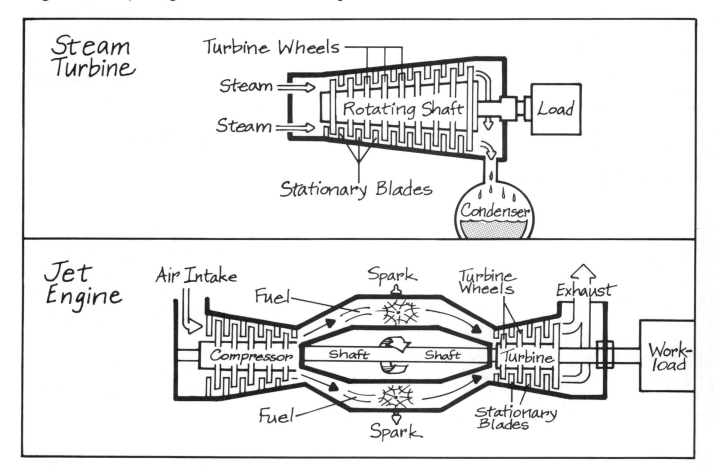

Questions

Discuss these questions with a partner and write responses in your science journal.

1. How does the steam-driven turbine work?
2. How does the jet engine work?
3. Which turbine consumes the most fuel?
4. Which turbine is most ecologically efficient?
5. How is each mechanism similar or different?
6. Why do you think a jet engine uses jet fuel instead of steam?

Extension Activities

Factory Design
(Literature/Art)

Discuss *Charlie and the Chocolate Factory* by Roald Dahl. Have students design and draw their own chocolate factories using a different form of energy to power equipment in each room. Remind them to include a source of power to open the gate! Students can write descriptive paragraphs describing their creations in detail.

Nuclear Waste
(Writing/Group Discussion/Research)

Nuclear waste is very dangerous because it undergoes radioactive decay although no longer capable of fission. Radioactivity produces high-energy radiation which can damage living cells, causing cancer and death. For safety, nuclear waste is coated with a sealing substance and buried in guarded facilities.

Have student groups research, discuss, and write solutions for nuclear waste storage. Older students can use periodicals to discover the present state of nuclear waste management. Students can also hold class debates on the following topics:

- Danger created by nuclear waste can be minimized by safer handling and storage.
- Nuclear energy is the best source of energy for countries who do not have ready access to fossil fuels.
- Nuclear reactors should be illegal.

Control Rods
(Research/Writing)

Concrete vaults surround nuclear reactor cores made of titanium or stainless steel. Within the core, 600°F pressurized water immerses uranium pellets and circulates through the boiler, creating steam. Long carbon rods, lowered into the water, absorb excess neutrons, moderating reactions and avoiding meltdowns. Nuclear reactions continue until most fissionable material splits into smaller nuclei. Have students choose an incident to research such as Three Mile Island or Chernobyl. Encourage them to respond to these questions: *What happened and why? How can it be avoided in the future? Is nuclear energy worth the risks?*

Waste Disposal
(Science Experiment)

Because nuclear waste continues to decay for hundreds of years, it is important to test containment procedures to prevent leakage into our environment. In this experiment, a pellet of sodium hydroxide (NaOH) represents a pellet of nuclear waste. Students will test the effectiveness of materials that prevent leakage.

> **Caution:** Do not touch sodium hydroxide pellets. NaOH causes chemical burns. Wear safety goggles and plastic gloves.

1. Fill four small beakers one-half full with water and add four drops of phenolphthalein indicator solution to each.
2. Using forceps, place one pellet of NaOH in one of the beakers.
3. Wrap another pellet in aluminum foil, one in clay, and one in plastic wrap secured with a twist tie.
4. Drop each of the wrapped pellets into one of the remaining beakers and cover them with plastic wrap.
5. Observe each day for three days to determine if there is leakage from the wrappings. Make a data chart and record your observations in your journal for each day.

Note: Clay is the best way to encase "nuclear waste" because it will not leak.

Culminating Activities

Energy Fair

Plan a culmination day for the energy unit. Brainstorm with students some creative activities for the day, such as:

- Invite cooperative groups to give oral presentations on energy topics. Have them design group posters depicting the main features and issues concerning their energy topic. Invite audiences (students, staff, parents) to listen and ask questions of teams and panels.

- Challenge student teams to debate which energy source is most practical for the future. Invite audiences to listen and provide feedback as to the effective use of debating skills.

- Invite students to design time capsules to be opened in 100 years. Time capsules should contain important information presently known about each energy category. Students should include recommendations to future generations about how to best benefit from and protect our current energy sources. Have time capsules on display for the Energy Fair.

- Ask students to write and illustrate their own "environmental concerns" books similar to *The Lorax* by Dr. Seuss. Encourage them to read their books to other classes. Bind these books into a class collection and display them at the Energy Fair.

- Challenge students to create an Energy Conservation Booth. Encourage students to provide pamphlets they have written and illustrated which describe simple ways to conserve energy.

- Invite a group of students to plan and cook creative refreshments using energy derived from the sun, wind, or water. This would be a fun place to integrate solar cookers and waterwheels.

Once students have finalized a list of creative activities, help them identify tasks and form cooperative teams. Provide time each day for groups to work on their projects. Use invitations and posters to advertise the special event. And, don't forget the video camera!

Career Corner

Interviews

Arrange a field trip to a local business to provide students with the opportunity to learn about careers in the energy field. Students might interview mechanics, plant managers, research scientists, and engineers (mechanical, electrical, nuclear, etc.). As an extension, have students write a job description for one career to post on a HELP WANTED bulletin board. Students can then conduct mock interviews by taking turns playing the role of the employer and the prospective employee.

Guest Speakers

Invite guest speakers from the community to share information with the class about their careers in the energy field. Include careers such as physicist, engineer, technology director, and industrial research scientist. Allow time for students to ask questions they have prepared beforehand.

Career Interview

Name: _____

Interviewee: _____

Career Role: _____

1. How long have you been in this career? _____

2. What are your main responsibilities? _____

3. Why is your job important? _____

4. What kind of training or education was necessary before you were able to begin your career? _____

5. What is your favorite and least favorite aspect of your job? _____

6. What are some interesting and/or important things you have learned during your time in this career? _____

Assessment

Ideally, science assessment includes evaluation of conceptual understanding, hands-on experiences, application of knowledge, and communication of learning. The following tools may assist you:

Science Portfolios—These are collections of representative student work which may include science journals, scientific method and processes sheets (pages 6–7), photos, and videos of special student projects. You may also wish to include anecdotal records, summaries of parent/teacher communication, and performance evaluations (page 61).

Anecdotal Records—Keep written records of observations which verify students' understanding of the scientific method and processes during hands-on activities. Use these to assist in student and parent conferencing.

Student Conferences—Ask students to discuss their most interesting experiment. Guide them using questions such as:

- Why did your experiment have those results?
- Did the results of your experiment support your hypothesis? Why or why not?
- What variables could have affected your results?
- What conclusions can you draw from the results?

Record student responses in writing or with a tape recorder.

Parent Conferencing—Discuss portfolios, student conferences, anecdotal records, and performance evaluations.

Performance Evaluations—Many teachers use science rubrics to evaluate individual student performance. Rubrics are paragraphs which describe benchmark levels of performance. The general performance evaluation on page 61 will help you give a numerical value for specific objectives addressed in many science rubrics. To obtain a rubric value, simply add the numbers and divide by 16. Or, create your own benchmark descriptions for the specific projects required in your classroom.

Experiment Rubric

Name: _____

Hypothesis: _____
_____

Description	Value
Exemplary performance on both the experiment and written evaluation. Detailed and careful attention given to the scientific method and data gathering. The conclusion demonstrates appropriate justification, insightful reasoning, and excellent critical thinking skills.	5
Experiment and written requirements completed correctly. Attention given to the steps of the scientific method. Able to support the conclusion with appropriate data.	4
Experiment complete but conclusion is not justified based upon data collected.	3
Experiment completed incorrectly.	2
Experiment incomplete.	1
No attempt made.	0

Name: _____

Performance Evaluation

Directions:
Circle the number which best reflects the frequency with which each behavior is observed.

0-never 1-rarely 2-occasionally 3-sometimes 4-often 5-frequently 6-always

Behavior							
Demonstrates understanding of important scientific concepts	0	1	2	3	4	5	6
Demonstrates understanding of the scientific method	0	1	2	3	4	5	6
Keeps accurate records of observations	0	1	2	3	4	5	6
Organizes data through categorizing and ordering	0	1	2	3	4	5	6
Draws logical conclusions from experiment results	0	1	2	3	4	5	6
Clearly communicates learning	0	1	2	3	4	5	6
Compares results of experiments with similar experiments	0	1	2	3	4	5	6
Relates prior knowledge to new learning	0	1	2	3	4	5	6
Makes connections across the curriculum	0	1	2	3	4	5	6
Makes inferences	0	1	2	3	4	5	6
Applies knowledge to solve problems	0	1	2	3	4	5	6
Demonstrates appropriate use of lab equipment	0	1	2	3	4	5	6
Demonstrates curiosity	0	1	2	3	4	5	6
Works cooperatively with others	0	1	2	3	4	5	6
Completes all work	0	1	2	3	4	5	6
Completes all work on time	0	1	2	3	4	5	6

Teacher Comments: _____

Student Comments: _____

Parent Comments: _____

Bibliography

Nonfiction

Asimov, Isaac. *How Did We Find Out About Solar Power?* Walker and Company, 1981. As part of a series on scientific discovery for grades 5–8, both the history and future of solar power are examined.

Bingham, Jane. *The Usborne Book of Science Experiments.* Usborne Publishing, 1991. Includes experiments in over 30 scientific topics.

Chaffin, Lillie D. *Coal Energy and Crisis.* Harvey House, 1974. Discusses coal as a source of energy and its future supply.

Coble, Charles. *A Look Inside Nuclear Energy.* Raintree Publishers, 1983. Traces the development of nuclear energy and examines related safety issues.

Cummings, Richard. *Make Your Own Alternative Energy.* David McKay Co., Inc., 1979. Explains solar power, water power, wind power, nuclear energy, integrated systems, and related projects.

Douglas, John H. *The Future World of Energy.* Walt Disney Productions, 1984. A well-documented look at the past, present, and future of energy.

Edom, Helen. *Science with Water.* Usborne Publishing, 1992. Includes motivating water power experiments and activities.

Epstein, Sam and Beryl. *The First Book of Electricity.* Franklin Watts, 1977. Illustrated, step-by-step electricity experiments.

Fuchs, Erich. *What Makes a Nuclear Power Plant Work?* Delacorte Press, 1972. Describes the workings of a nuclear power plant using detailed pictures and simple text.

Gutnik, Martin J. *Electricity: From Faraday to Solar Generators.* Franklin Watts, 1986. The history of electricity from early Greeks to modern day.

Harrison, George Russell. *First Book of Energy.* Franklin Watts, 1965. A thorough treatment of energy, including math concepts and illustrations.

Jennings, Terry. *Energy and Forces.* Smithmark Publishers, 1992. Over 30 easy energy experiments to do at home.

Kaplan, Sheila. *Solar Energy.* Raintree Publishers, 1983. A thorough look at the sun's energy and the development of solar cells and solar-collecting platforms.

Lafferty, Peter. *Force and Motion* (Eyewitness Science Series). Dorling Kindersley, 1992. Excellent illustrations, explanations, and basic history of discoveries related to force and motion.

Landt, Dennis. *Catch the Wind.* Four Winds Press, 1976. Traces the use of wind power, as used by navigators and channeled through windmills.

Leon, George De Lucenay. *The Electricity Story.* Arco, 1983. The history of electricity and experiments that played a historical role.

Macauley, David. *The Way Things Work.* Houghton Mifflin, 1988. Fascinating illustrations and explanations of all types of machinery.

Math, Irwin. *More Wires & Watts.* Macmillan, 1988. Experiments and projects that demonstrate the fundamentals of electricity and magnetism.

McDonald, Lucile. *Windmills: An Old-New Energy Source.* Elswer-Dutton Publishing Co., 1981. History of windmills and windmills in art.

Metos, Thomas H. and Gary G. Bitter. *Exploring with Solar Energy.* Julian Messner, 1978. Excellent experiments, historical drawings, and photographs.

Notkin, Jerome J., Ed.D. *The How and Why Wonder Book of Electricity.* Grossett and Dunlap, 1960. Simple explanations of energy experiments told in story form.

Petersen, David. *Solar Energy at Work.* Childrens Press, 1985. Brief descriptions of past and present uses of solar energy.

Rickard, Graham. *Wind Energy.* Gareth Stevens Children's Books, 1991. A discussion on alternative energy, including directions for making a wind turbine.

Shuttleworth, John. *The Handbook of Homemade Power.* Bantam, 1974. Homespun machines and energy applications, including Mother Earth articles.

Sootin, Harry. *Experiments with Electric Currents.* W.W. Norton and Company, Inc., 1969. An in-depth study weaving history and experiments together.

Sterland, E.G. *Energy into Power: The Story of Man and Machines.* The American Museum of Natural History, 1967. Time line with photos and drawings of scientists.

Ward, Alan. *Experimenting with Batteries, Bulbs, and Wires.* Chelsea Juniors, 1991. Hands-on activities using batteries, bulbs, and wires.

Zubrowski, Bernie. *Blinkers and Buzzers: Building and Experimenting with Electricity.* Morrow, 1991. Electricity and magnetism experiments and projects.

Fiction

Bourne, Miriam Anne. *Bright Lights to See By*. Coward, McCann & Geoghegan, 1975. Rival hotel owners argue the value of electric lights and a play performance puts their debate to a test.

Dahl, Roald. *Charlie and the Chocolate Factory*. Puffin books, 1988. The adventures of a boy who wins a tour of Willy Wonka's wild and crazy chocolate factory.

DeWeese, Gene. *The Calvin Nullifier*. McGraw-Hill, 1987. Calvin Willeford has a special aura that makes electrical devices behave strangely. When a space probe begins to malfunction, Calvin discovers a connection between himself and these phenomena.

Firmin, Peter. *Basil Brush and the Windmills*. Englewood Cliffs, 1980. With one penny left to pay utility bills, Basil Brush and his friend decide to build a windmill for power.

Hadley, Eric and Tessa. *Legends of the Sun and Moon*. Cambridge University Press, 1989. A multicultural collection of sun and moon tales.

MacGregor, Ellen and Dora Pantell. *Miss Pickerell and the Geiger Counter*. McGraw-Hill, 1953. Miss Pickerell discovers the secrets of uranium and radiation and points an imaginative finger to the future.

MacGregor, Ellen and Dora Pantell. *Miss Pickerell Tackles the Energy Crisis*. McGraw-Hill, 1980. Miss Pickerell, a spry, resourceful, humorous farmer, involves herself in a science challenge.

McGovern, Ann. *Aesop's Fables*. Scholastic, 1963. A collection of fables, including sun, water, and wind themes.

Osborne, Mary Pope. *Favorite Greek Myths*. Scholastic, 1989. A delightful collection of Greek myths.

Rice, James. *Gason Drills an Offshore Oil Well*. Pelican, 1982. Gason drills an offshore oil well and invests his money to help the energy crisis. An authentic replication of the industry and a basic introduction to our energy needs.

Seuss, Dr. *The Lorax*. Random House, 1971. Imaginary creatures deplete natural resources in their environment to satisfy their greed. Told in rhyme.

Townsend, Tom. *Queen of the Wind*. Panda Books, 1989. 19-year-old Blaze finds adventure when she takes a job as an instructor at a sailing school in Texas.

Verne, Jules. *Around the World in 80 Days*. Dodd/Mead, 1956. Various forms of energy and vehicles transport Phineas Fogg and his manservant, Passepartout, around the globe in 80 days.

Verne, Jules. *20,000 Leagues Under the Sea*. Nelson Doubleday, 1956. Captain Nemo seeks to rule the world as he travels in his electrically-driven submarine.

Energy Resources

Technology

Laser Disc

Windows on Science (series)
Physical Science, Vol. II
Optican Data Corp., 1989.

Video

Nuclear Energy/Nuclear Waste
Schlessinger Video Productions, 1993.
The Children's Video Encyclopedia
Vol. XVII: Science, Sound, and Energy
Segments: "What is Energy?"
"What is Fuel?"
Tell My Why, Inc., 1989.

Free Catalogs and Low-cost Materials

Energy (Subject: Bibliography 303)
U.S. Government Printing Office
Superintendent of Documents
Washington, DC 20401
FAX 1-202-512-1716

This free bibliography lists books and subscriptions available at a low cost.

Energy Information Administration (nuclear)
Energy End Use and Integrated Statistics
1000 Independence Avenue, S.W.
Washington, DC 20585

Maintains data on energy consumption in the residential, commercial, industrial, and transportation sectors.

Nuclear Regulatory Commission (nuclear)
1-800-368-5642

Free publications on nuclear energy for teachers and students.

Renewable Energy Conservation Information
Department of Energy
1-800-523-2929

Provides resources for teachers and hands-on kits for students.

Glossary of Energy Terms

anemometer—a tool for measuring wind velocity

cogeneration—the production of electricity using waste heat (e.g., steam) from an industrial process

dams—barriers that hold back water, often forming reservoirs or lakes

distillation—the purification of liquid through evaporation and condensation processes

electromagnetic spectrum—radio waves, infrared waves, visible waves, visible light, ultraviolet light, microwaves, X rays, and gamma waves

external combustion engine—a heat engine (e.g., steam engine) that gets its heat from fuel consumed outside the cylinder

fission—the breaking up of large nuclei into smaller nuclei

fossil fuels—fossilized plant and animal matter found as coal, crude oil, and natural gas

fusion—the buildup of small nuclei into larger nuclei

galvanometer—a tool for detecting small amounts of electric current

generator—a machine that changes mechanical energy into electrical energy

half-life—the time it takes for half of a particular substance to undergo a process

hydraulic machines—mechanical devices that use water as a source of energy

hydroelectric power sources—dams, waterfalls, waves, and tidal currents

internal combustion engine—a heat engine in which the combustion that generates the heat occurs inside rather than outside the engine

nuclear energy—energy released during a nuclear reaction such as fission or fusion

nuclear reactors—a power plant in which the controlled release of nuclear energy takes place

nuclei—plural of nucleus

nucleus—the small, dense center of an atom where protons and neutrons are located

petroleum—crude oil

power plants—installations that convert stored solar energy into electricity

radioactivity—the property of some elements (e.g., uranium) or isotopes (e.g., carbon 14) when they are continuing to emit rays due to their nuclei disintegrating

refined—freed from impurities; fractioned (e.g., crude oil refined into useful compounds such as gasoline, heating oils, and lubricants)

renewable energy—energy that cannot be depleted such as wind, water, and solar

solar energy—radiant energy from the sun

spent—when an element is no longer useful in a nuclear reactor

stable element—an element that remains in its current atomic state

turbine—a rotary engine run by a reaction and/or impulse of a current of fluid such as wind, steam, or water

unstable element—an element continuing to break down rather than remain in its current atomic state; readily changing in atomic composition

wind—air movement; an effect of solar energy

wind prospectors—individuals who search for locations that experience consistent, strong wind flow

windmill—a machine for harnessing wind energy